Selected Titles in This Series

602 **Igor Fulman,** Crossed products of von Neumann algebras by equivalence relations and their subalgebras, 1997

601 **Jack E. Graver and Mark E. Watkins,** Locally finite, planar, edge-transitive graphs, 1997

600 **Ambar Sengupta,** Gauge theory on compact surfaces, 1997

599 **Tai-Ping Liu and Yanni Zeng,** Large time behavior of solutions for general quasilinear hyperbolic-parabolic systems of conservation laws, 1997

598 **Valentina Barucci, David E. Dobbs, and Marco Fontana,** Maximality properties in numerical semigroups and applications to one-dimensional analytically irreducible local domains, 1997

597 **Ragnar-Olaf Buchweitz and John J. Millson,** CR-geometry and deformations of isolated singularities, 1997

596 **Paul S. Bourdon and Joel H. Shapiro,** Cyclic phenomena for composition operators, 1997

595 **Eldar Straume,** Compact connected Lie transformation groups on spheres with low cohomogeneity, II, 1997

594 **Solomon Friedberg and Hervé Jacquet,** The fundamental lemma for the Shalika subgroup of $GL(4)$, 1996

593 **Ajit Iqbal Singh,** Completely positive hypergroup actions, 1996

592 **P. Kirk and E. Klassen,** Analytic deformations of the spectrum of a family of Dirac operators on an odd-dimensional manifold with boundary, 1996

591 **Edward Cline, Brian Parshall, and Leonard Scott,** Stratifying endomorphism algebras, 1996

590 **Chris Jantzen,** Degenerate principal series for symplectic and odd-orthogonal groups, 1996

589 **James Damon,** Higher multiplicities and almost free divisors and complete intersections, 1996

588 **Dihua Jiang,** Degree 16 Standard L-function of $GSp(2) \times GSp(2)$, 1996

587 **Stéphane Jaffard and Yves Meyer,** Wavelet methods for pointwise regularity and local oscillations of functions, 1996

586 **Siegfried Echterhoff,** Crossed products with continuous trace, 1996

585 **Gilles Pisier,** The operator Hilbert space OH, complex interpolation and tensor norms, 1996

584 **Wayne W. Barrett, Charles R. Johnson, and Raphael Loewy,** The real positive definite completion problem: Cycle completability, 1996

583 **Jin Nakagawa,** Orders of a quartic field, 1996

582 **Darryl McCollough and Andy Miller,** Symmetric automorphisms of free products, 1996

581 **Martin U. Schmidt,** Integrable systems and Riemann surfaces of infinite genus, 1996

580 **Martin W. Liebeck and Gary M. Seitz,** Reductive subgroups of exceptional algebraic groups, 1996

579 **Samuel Kaplan,** Lebesgue theory in the bidual of $C(X)$, 1996

578 **Ale Jan Homburg,** Global aspects of homoclinic bifurcations of vector fields, 1996

577 **Freddy Dumortier and Robert Roussarie,** Canard cycles and center manifolds, 1996

576 **Grahame Bennett,** Factorizing the classical inequalities, 1996

575 **Dieter Heppel, Idun Reiten, and Sverre O. Smalø,** Tilting in Abelian categories and quasitilted algebras, 1996

574 **Michael Field,** Symmetry breaking for compact Lie groups, 1996

573 **Wayne Aitken,** An arithmetic Riemann-Roch theorem for singular arithmetic surfaces, 1996

572 **Ole H. Hald and Joyce R. McLaughlin,** Inverse nodal problems: Finding the potential from nodal lines, 1996

(See the AMS catalog for earlier titles)

Crossed Products of von Neumann Algebras by Equivalence Relations and Their Subalgebras

MEMOIRS
of the
American Mathematical Society

Number 602

Crossed Products of von Neumann Algebras by Equivalence Relations and Their Subalgebras

Igor Fulman

March 1997 • Volume 126 • Number 602 (third of 5 numbers) • ISSN 0065-9266

American Mathematical Society
Providence, Rhode Island

1991 *Mathematics Subject Classification.*
Primary 47D25, 46L10, 47A20.

Library of Congress Cataloging-in-Publication Data

Fulman, Igor, 1965–
Crossed products of von Neumann algebras by equivalence relations and their subalgebras / Igor Fulman.
p. cm. — (Memoirs of the American Mathematical Society, ISSN 0065-9266 ; no. 602)
"March 1997, volume 126, number 602 (third of 5 numbers)."
Includes bibliographical references.
ISBN 0-8218-0557-6 (alk. paper)
1. Von Neumann algebras—Crossed products. 2. Equivalence relations (Set theory. I. Title. II. Series.
QA3.A57 no. 602
[QA326]
510 s—dc21
[512′.55]
96-47955
CIP

Memoirs of the American Mathematical Society

This journal is devoted entirely to research in pure and applied mathematics.

Subscription information. The 1997 subscription begins with number 595 and consists of six mailings, each containing one or more numbers. Subscription prices for 1997 are $414 list, $331 institutional member. A late charge of 10% of the subscription price will be imposed on orders received from nonmembers after January 1 of the subscription year. Subscribers outside the United States and India must pay a postage surcharge of $30; subscribers in India must pay a postage surcharge of $43. Expedited delivery to destinations in North America $35; elsewhere $110. Each number may be ordered separately; *please specify number* when ordering an individual number. For prices and titles of recently released numbers, see the New Publications sections of the *Notices of the American Mathematical Society.*

Back number information. For back issues see the *AMS Catalog of Publications.*

Subscriptions and orders should be addressed to the American Mathematical Society, P. O. Box 5904, Boston, MA 02206-5904. *All orders must be accompanied by payment.* Other correspondence should be addressed to Box 6248, Providence, RI 02940-6248.

Memoirs of the American Mathematical Society is published bimonthly (each volume consisting usually of more than one number) by the American Mathematical Society at 201 Charles Street, Providence, RI 02904-2294. Periodicals postage paid at Providence, RI. Postmaster: Send address changes to Memoirs, American Mathematical Society, P. O. Box 6248, Providence, RI 02940-6248.

This publication is indexed in *Science Citation Index*®, *SciSearch*®, *Research Alert*®, *CompuMath Citation Index*®, *Current Contents*®/*Physical, Chemical & Earth Sciences.*
Printed in the United States of America.

∞ The paper used in this book is acid-free and falls within the guidelines established to ensure permanence and durability.

10 9 8 7 6 5 4 3 2 1 01 00 99 98 97

CONTENTS

ABSTRACT

In this work the author introduces and studies the construction of the crossed product of a von Neumann algebra M decomposed into the direct integral $M = \int_X M(x)d\mu(x)$ by an equivalence relation on X with countable cosets.

This construction is the generalization of the construction of the crossed product of an abelian von Neumann algebra by an equivalence relation introduced by J. Feldman and C. C. Moore in [15].

Many properties of this construction are studied. In particular, the structure theorem generalizing Theorem 1 in [15] is proved. The generalizations of the Spectral Theorem on Bimodules (see [25, Theorem 2.5]) and of the theorem on dilations (see [26, Theorem 1]) are proved too.

1991 *Mathematics Subject Classification.* 47D25, 46L10, 47A20.

Key words and phrases. Von Neumann algebra, crossed product, equivalence relation, bimodule, subalgebra, dilation.

CHAPTER 1

INTRODUCTION

In the papers [14, 15] J. Feldman and C. C. Moore have introduced and studied the construction which can be named the "crossed product" of an abelian von Neumann algebra by an equivalence relation. This construction is analogous to the crossed product of a von Neumann algebra M by a group (see the example in Subsection 20.1 below).

Furthermore, J. Feldman and C. C. Moore have developed the 2-cohomology theory for equivalence relations. This theory allowed them to prove the structure theorem on "crossed products" (see [15, Theorem 1]), i. e. to give the necessary and sufficient condition for a von Neumann algebra to be a "crossed product" of some abelian subalgebra by some equivalence relation.

J. Feldman and C. C. Moore have also studied the automorphisms and isomorphisms of such algebras.

The equivalence relation in the construction of Feldman and Moore provides the algebra with coordinates; i. e. allows the elements of the algebra to be represented as "generalized complex matrices". These coordinates were used by Feldman and Moore to study the automorphisms of these algebras.

Subalgebras (in general non selfadjoint) of these algebras were studied by P. Muhly, K.-S. Saito and B. Solel [25]. Such subalgebras turned out to be associated with partial orders. A key result in the study of these subalgebras is the Spectral theorem on bimodules (see [25, Theorem 2.5]) which allows one to associate bimodules over the diagonal M with subsets of the equivalence relation.

Using coordinates for the study of operator algebras has proved very useful in several classes of algebras: e. g. CLS algebras (see [3]) and subalgebras of groupoid C^*-algebras (see [27]; in particular, for triangular AF algebras see [31]).

In the paper [26] the authors proved a theorem on dilations of a representation of a subalgebra of such algebra to a representation of the whole algebra (see [26, Theorem 1]).

In the present paper we introduce and study the construction which is a generalization of the construction of J. Feldman and C. C. Moore to the case where the von Neumann algebra M is not commutative, but it is decomposed into the direct integral $M = \int_X M(x) d\mu(x)$. This integral may be the central decomposition (where all $M(x)$ are factors), but it is not necessary.

For this case we define and study the crossed product of a von Neumann algebra by an equivalence relation. We define also the analogue of the 2-cohomology theory in our case. This allows us to prove the structure theorem (see Theorem 12.2 below) which is the exact analogue of the structure theorem of J. Feldman and C. C. Moore.

Received by the editor October 22, 1994.

We, thus, obtain a class of algebras equipped with coordinates (with "non commutative fibers"). This allows us to represent the elements of such an algebra as "generalized operator matrices". In Sections 10, 11 and 13 we use these coordinates to study the automorphisms and isomorphisms of these von Neumann algebras. We are also able to describe the analytic algebras associated with flows of automorphisms (see Section 16).

In our case, under the condition that the equivalence relation is hyperfinite, we also prove the spectral theorem on bimodules (see Theorem 15.18 below). As in the paper [25], we use this theorem to describe subalgebras of the crossed product algebras (in particular, non-selfadjoint subalgebras).

In Section 18 we study the case where the equivalence relation is hyperfinite. We prove that, in this case, every contractive σ-weakly continuous representation of a σ-Dirichlet subalgebra is completely contractive, hence has a dilation to a representation of the von Neumann algebra. In the Feldman-Moore setting this was proved by Muhly and Solel in [26]. Such a dilation property was used in the study of representations of non selfadjoint algebras (see [1, 26, 28, 11]).

Finally, we show the interrelations of our construction with the construction of the crossed product of a von Neumann algebra by a groupoid, introduced by Takehiko Yamanouchi in his paper [37]. His construction is different from ours, and he studied not only actions of groupoids, but also the dual coactions. Nevertheless, it turns to be that when our case and the case of Yamanouchi "intersect", the construction of Yamanouchi in essential coincides with our construction.

We wish to note that our algebras are also a special case of algebras obtained by Fell bundles over measured equivalence relations (see [36]).

The author expresses his sincere graditude to his scientific superviser, Professor Baruch Solel, for his helpful guidance and constant encouragement. The author also wish to thank Professor P. Muhly for several helpful discussions.

CHAPTER 2

PRELIMINARIES

Let M be a von Neumann algebra acting on a Hilbert space H. Let M be decomposed into a direct integral $M = \int_X M(x)d\mu(x)$ where $(X, \mathcal{B}, \mu)$ is a standard Borel measure space. Let $H = \int_X H(x)d\mu(x)$ be the corresponding decomposition of H. Let M possess a cyclic and separating vector $\phi_0 = \int_X \phi_0(x)d\mu(x)$.

Our goal is to develop the Feldman-Moore construction [14, 15] for M in place of $L^\infty(X, \mathcal{B}, \mu)$.

Let R be an equivalence relation on X. We will write $x \sim y$ for $(x, y) \in R$. Let R be Borel as a subset of $X \times X$. Let $\triangle = \{(x, x) \mid x \in X\}$; then $\triangle \subset R$.

Let $\pi_r : R \to X$, $\pi_l : R \to X$, $\theta : R \to R$ be defined as follows:

$$\pi_r(x, y) = y, \qquad \pi_l(x, y) = x, \qquad \theta(x, y) = (y, x)$$

Let R be countable in the following sense: every coset of R is countable. Let ν be a measure on R defined by:

$$\nu(C) = \int_X \mathrm{card}(C \cap \pi_r^{-1}(y))d\mu(y)$$

for C Borel as a subset of $X \times X$.

This definition is correct by [14, Theorem 2], and

$$\int_R f(x, y)d\nu(x, y) = \int_X \sum_{x \sim y} f(x, y)d\mu(y)$$

Let Σ be the set of all Borel isomorphisms of X with graphs in R, Σ^p be the set of all partial Borel isomorphisms of X with graphs in R.

Let $\tilde{\Gamma}$ be the collection of all maps $f : R \to \bigcup_{y \in X} H(y)$ such that:

1. for every $(x, y) \in R$: $f(x, y) \in H(y)$;

2. for every $\tau \in \Sigma$ the vector field $x \mapsto f(\tau x, x)$ belongs to H.

Let for $(x, y) \in R$: $H(x, y) \equiv H(y)$.

Then $(\{H(x, y)\}_{(x,y) \in R}, \tilde{\Gamma})$ is a ν-measurable field of Hilbert spaces in the sense of Dixmier [10].

By [10], the field $(\{H(x, y)\}_{(x,y) \in R}, \tilde{\Gamma})$ defines the direct integral of Hilbert spaces:

$$\tilde{H} = \int_R H(x, y)d\nu(x, y)$$

NOTATION. Let $\alpha = \{\alpha_{(x,y)}\}_{(x,y) \in R}$ be a collection of isomorphisms of von Neumann algebras such that:

1. for every $(x,y) \in R$, $\alpha_{(x,y)}$ is an isomorphism from $M(y)$ onto $M(x)$;
2. for $x \sim y \sim z$: $\alpha_{(x,z)} = \alpha_{(x,y)} \circ \alpha_{(y,z)}$;
3. if $a = \int_X a(x)d\mu(x) \in M$, $\tau \in \Sigma$, then $x \mapsto \alpha_{(x,\tau x)}(a(\tau x))$ is a measurable operator field (i. e. there exist $b \in M$ such that $b(x) = \alpha_{(x,\tau x)}(a(\tau x))$ a. e.).

Let $R^2 = \{(x,y,z) \mid x,y,z \in X, x \sim y \sim z\}$.
Let $c : R^2 \to \bigcup_{z\in X} B(H(z))$ be a map with the following properties:

1. for every $(x,y,z) \in R^2$, $c(x,y,z)$ is a unitary operator on $H(z)$;
2. for every $(x,y,z) \in R^2$, $c(x,y,z)$ normalises $M(z)$, i. e.
$$c(x,y,z)M(z)c(x,y,z)^{-1} = M(z)$$
3. for a. e. $x \in X$ and for every $y \sim z \sim v \sim w \sim x$:
$$c(x,y,z)c(y,v,z)c(x,v,z)^{-1} \in M(z)$$
and
$$c(x,y,z)c(y,v,z)c(x,v,z)^{-1} = \alpha_{(z,w)}(c(x,y,w)c(y,v,w)c(x,v,w)^{-1})$$
4. for a. e. $x \in X$ and for $y \sim z \sim v \sim x$:
$$\alpha_{(z,v)}(c(x,y,v)^{-1})c(x,y,z) \in M(z)'$$

REMARK. If ϕ_0 is a separating and cyclic vector for M, then for a. e. $y \in X$, $\phi_0(y)$ is a separable and cyclic vector for $M(y)$, so for a. e. $(x,y) \in R$, $\alpha_{(x,y)}$ is spatial, so it can be expanded to all of $B(H(y))$.

5. for every $(x,y,z) \in R^2$: $c(x,x,y) = 1_{H(y)}$, and $c(x,y,z)^{-1} = c(y,x,z)$;
6. for every $\tau, \sigma \in \Sigma$, the map $x \mapsto c(\tau x, \sigma x, x)$ is Borel, i. e. there exists $a \in \int_X B(H(x))d\mu(x)$ such that $a(x) = c(\tau x, \sigma x, x)$ a. e.

For every $a \in M$, $\tau \in \Sigma$ we define:
$$\begin{aligned}(I(a)f)(x,y) &= \alpha_{(y,x)}(a(x))f(x,y), & f \in \tilde{H}\\ (\pi(\tau)f)(x,y) &= c(x,\tau x,y)f(\tau x,y), & f \in \tilde{H}\end{aligned}$$
where $a = \int_X a(x)d\mu(x)$ is the decomposition of a.

Then because of the measurability of α and c: $I(a)f \in \tilde{H}$, $\pi(\tau)f \in \tilde{H}$, and further: $I(a) \in B(\tilde{H})$, $\pi(\tau) \in B(\tilde{H})$ for every such a and τ. Moreover, for every $\tau \in \Sigma$ the operator $\pi(\tau)$ is unitary.

Definition 2.1 *The von Neumann algebra $\tilde{M}$ generated by $I(a)$ and $\pi(\tau)$ for all $a \in M$, $\tau \in \Sigma$ will be called a* **crossed product** *of the von Neumann algebra M by the equivalence relation R. We will write:*
$$\tilde{M} = M \bowtie R$$

CHAPTER 3

UNITARY REALIZATION OF $\alpha_{(y,x)}$

Let $\phi_0 \in H$ be a separating and cyclic vector for M, $\tilde{\phi}_0(x,y) = \delta_{(x,y)}\phi_0(y)$.

NOTATION. Let for $\tau \in \Sigma$, $a \in M$: $\alpha_\tau(a)(x) = \alpha_{(x,\tau^{-1}x)}(a(\tau^{-1}x))$.

Then $x \mapsto \alpha_\tau(a)(x)$ is measurable, i. e. there exists

$$\alpha_\tau(a) = \int_X \alpha_\tau(a)(x) d\mu(x) \in M$$

It is clear that α_τ is a *-automorphism of M. It is not hard to see that $\alpha_{\tau_1} \circ \alpha_{\tau_2} = \alpha_{\tau_1\tau_2}$ for $(\tau_1, \tau_2 \in \Sigma)$.

Let $d\nu^{-1}(x,y) = d\nu(y,x)$ and $D(x,y) = (d\nu^{-1}/d\nu)(x,y)$. It is known (see [14, Corollary 2 on page 294]) that for a. e. $x \sim y \sim z$: $D(x,y)D(x,z) = D(x,z)$.

By the general theorem [6, Corollary 2.5.32], for $\tau \in \Sigma$ there exist unitary operators U_τ in $B(H)$ such that $\alpha_\tau(a) = U_\tau a U_\tau^*$, $U_{\tau_1}U_{\tau_2} = U_{\tau_1\tau_2}$ (hence $U_{\tau^{-1}} = U_\tau^*$) and $U_\tau J = J U_\tau$, where J is the unitary involution from the Tomita-Takesaki theory, associated with M and ϕ_0.

By [14, Proposition 2.2], for $\phi \in \Sigma$ we have: $(d\phi_*\mu/d\mu)(y) = D(\phi^{-1}y, y)$ for a. e. $x \in X$. Thus, $d\mu(\phi^{-1}y) = d\phi_*\mu(y) = D(\phi^{-1}y, y)d\mu(y)$. Take $\phi = \tau^{-1}$ and get :

$$d\mu(\tau y) = D(\tau y, y) d\mu(y)$$

Moreover, for $a \in A$: $U_\tau a U_\tau^* = \alpha_\tau(a) = a \circ \tau^{-1}$. By [16, Proposition 4.1], this implies that there exist unitary operators U_x^τ from $H(x)$ onto $H(\tau x)$ defined a. e. such that $(U_\tau f)(x) = D(x, \tau^{-1}x)^{-1/2} U^\tau_{\tau^{-1}x} f(\tau^{-1}x)$ for a. e. $x \in X$, $f \in H$.

NOTATION. Let $U_{(x,\tau^{-1}x)} = U^\tau_{\tau^{-1}x}$.

The notation above is well defined for a. e. $x \in X$, i. e. it depends only on $\tau^{-1}x$, not on $\tau \in \Sigma$.

REMARK. By the definition, for $f \in H$, a. e. $x \in X$:

$$U_{(x,\tau^{-1}x)} f(\tau^{-1}x) = D(x, \tau^{-1}x)^{1/2} (U_\tau f)(x)$$

It can be checked immediately that for a. e. $x \in X$, $y \sim z \sim x$, $a \in M(x)$:

1. $U_{(y,x)} = U_{(y,z)}U_{(z,x)}$;
2. $\alpha_{(y,x)}(a) = U_{(y,x)} a U_{(x,y)}$;
3. $J(y)U_{(y,x)} = U_{(y,x)}J(x)$.

Corollary 3.1 $\alpha_{(x,y)}(J(y)) = J(x)$ *a. e.*

(See the Remark on $\alpha_{(x,y)}$ at the end of Section 2.)

CHAPTER 4

CONSTRUCTION OF $\tilde{M}^{\nabla}$

NOTATION. Let for every $a' \in M'$, $\tau \in \Sigma$, $f \in \tilde{H}$:

$$\begin{aligned}(I'(a')f)(x,y) &= c(x,y,y)a'(y)c(x,y,y)^{-1}f(x,y)\\ (\pi'(\tau)f)(x,y) &= D(\tau^{-1}y,y)^{1/2}c(x,\tau^{-1}y,y)U_{(y,\tau^{-1}y)}\\ &\qquad\qquad c(x,\tau^{-1}y,\tau^{-1}y)^{-1}f(x,\tau^{-1}y)\end{aligned}$$

Then $I'(a'), \pi'(\tau) \in B(H)$. Moreover, $\pi'(\tau)$ is unitary.

NOTATION. Let $\tilde{M}^{\nabla}$ be the von Neumann algebra generated by $I'(a')$ and $\pi(\tau)$ for all $a' \in M'$, $\tau \in \Sigma$.

Theorem 4.1 *Every operator in $\tilde{M}^{\nabla}$ commutes with every operator in $\tilde{M}$.*

PROOF. It is enough to check that for $a \in M$, $a' \in M'$, $\tau, \sigma \in \Sigma$: $I(a)$ and $\pi(\tau)$ commute with $I'(a')$ and $\pi'(\sigma)$.

It can be checked by the straight calculation.

□

CHAPTER 5

COORDINATE REPRESENTATION OF ELEMENTS OF $\tilde{M}$

In this section for every $T \in \tilde{M}$ we define some function $T(\cdot,\cdot)$ from R to $\bigcup_{y\in X} M(y)$ whose properties are analogous to the properties of the function $a(T)$ in [15]. So, let $T \in \tilde{M}$.

NOTATION. For every $\phi \in \Sigma$ and for all $f \in H$ we define:

$$\tilde{f}_\phi(x,y) = \begin{cases} 0 & \text{if } x \neq y \\ c(\phi y, y, y)^{-1} f(y) & \text{if } x = y \end{cases}$$

Then $\tilde{f}_\phi = \int_R \tilde{f}_\phi(x,y) d\nu(x,y)$ belongs to $\tilde{H}$. So, $T\tilde{f}_\phi \in \tilde{H}$.

NOTATION. Let us define $(T_\phi f)(y) = (T\tilde{f}_\phi)(\phi y, y)$.

Then $T_\phi f = \int_X (T_\phi f)(y) d\mu(y)$ belongs to H. So, T_ϕ is an operator on H. It is bounded and $\|T_\phi\| \leq \|T\|$.

Let $T \in \tilde{M}$. It is easy to see that $T_\phi \in M$. Let $T_\phi = \int_X T_\phi(y) d\mu(y)$ be the decomposition of the operator T_ϕ.

NOTATION. Let for $\phi \in \Sigma$, $y \in X$: $T(\phi y, y) = T_\phi(y)$.

The notation above is well defined for a. e. $y \in X$, i. e. depends only on ϕy, not on $\phi \in \Sigma$.

Proposition 5.1 *For $a \in M$, $\tau \in \Sigma$, a. e. $(x,y) \in R$:*

$$(I(a))(x,y) = \begin{cases} 0 & \text{if } x \neq y \\ a(x) & \text{if } x = y \end{cases}, \qquad (\pi(\tau))(x,y) = \begin{cases} 0 & \text{if } y \neq \tau x \\ 1_{H(y)} & \text{if } y = \tau x \end{cases}$$

PROOF. The proof is straightforward.

□

Theorem 5.2 (See [15, Proposition 2.6].) *Let $f \in \tilde{H}$, $T, T_1, T_2 \in \tilde{M}$. Then for a. e. $(x,y) \in R$:*

$$(Tf)(x,y) = \sum_{z\sim y} \alpha_{(y,z)}(T(x,z)) c(x,z,y) f(z,y) \tag{5.1}$$

$$(T_1T_2)(x,y) = \sum_{z\sim y} \alpha_{(y,z)}(T_1(x,z)) c(x,z,y) T_2(z,y) c(z,y,y) c(x,y,y)^{-1} \tag{5.2}$$

where the series in (5.1) is understood with respect to the norm topology in $H(y)$, the series in (5.2) — with respect to the strong operator topology in $M(y)$.

PROOF.

1. (a) For a vector $f \in \tilde{H}$ such that $\operatorname{supp} f \subseteq \Gamma(\tau)$ with $\tau \in Sigma$, the verifying is straightforward.
   (b) Let $f \in \tilde{H}_f$, where $\tilde{H}_f = \{f \in \tilde{H} \mid \operatorname{supp} f \subset \sum_{i=1}^n \Gamma(\tau_i),\ \tau_i \in \Sigma,\ i = 1, \ldots, n\}$. Then $f = \sum_{i=1}^n f_i$, $\operatorname{supp} f \subset \Gamma(\tau_i)$, $i = 1, \ldots, n$. The equality (5.1) holds for every f_i. But $f(x, y) = \sum_{i=1}^n f_i(x, y)$, so (5.1) holds for $f \in \tilde{H}_f$.
   (c) Let $f \in \tilde{H}$ be arbitrary. Let $(\tau_i) \in \Sigma$ be a sequence such that $R = \bigcup_{i=1}^\infty \Gamma(\tau_i)$. Define $f_n \in \tilde{H}$ by

$$f_n(x, y) = \begin{cases} f(x, y) & \text{if there exists } i \leq n \text{ such that } (x, y) \in \Gamma(\tau_i) \\ 0 & \text{otherwise} \end{cases}$$

   Then $f_n \in \tilde{H}_f$, $f_n \to f$. Hence, $Tf_n \to Tf$. So by [10, Ch. II, §1, n. 5, Proposition 5(ii)] there exists a subsequence converging to Tf almost everywhere. Thus, we can pass to this sequence and assume that $Tf_n \to Tf$ a. e.
   By part (1b) of this proof:

$$(Tf_n)(x, y) = \sum_{z \sim y} \alpha_{(y,z)}(T(x, z))c(x, z, y)f_n(z, y)$$

   The left hand side converges to $(Tf)(x, y)$ almost everywhere. So, the right hand side converges too. But, by the choice of f_n, the right hand side is nothing but the partial sum of the series

$$\sum_{z \sim y} \alpha_{(y,z)}(T(x, z))c(x, z, y)f(z, y)$$

   Thus, this series converges to $(Tf)(x, y)$ a. e.

2. Let $f \in \tilde{H}$, $\operatorname{supp} f \subset \triangle$. Applying twice the argument of part (1) of this proof, we get that for a. e. $(x, y) \in R$:

$$\begin{aligned}(T_1T_2f)(x, y) &= (T_1(T_2f))(x, y) \\ &= \textstyle\sum_z \alpha_{(y,z)}(T_1(x, z))c(x, z, y)T_2(z, y)c(z, y, y)f(y, y)\end{aligned}$$

   On the other hand, by part (1) of this proof:

$$\begin{aligned}(Tf_n)(x, y) &= \sum_w \alpha_{(y,w)}((T_1T_2)(x, w))c(x, w, y)f(w, y) \\ &= (T_1T_2)(x, y)c(x, y, y)f(y, y)\end{aligned}$$

   This is true for every f as above. Hence:

$$(T_1T_2)(x, y)c(x, y, y) = \sum_z \alpha_{(y,z)}(T_1(x, z))c(x, z, y)T_2(z, y)c(z, y, y)$$

   where the series is understood as the strong limit. Finally,

$$(T_1T_2)(x, y) = \sum_z \alpha_{(y,z)}(T_1(x, z))c(x, z, y)T_2(z, y)c(z, y, y)c(x, y, y)^{-1}$$

□

Theorem 5.3 *For $T \in \tilde{M}$, for a. e. $(x,y) \in R$:*

$$T^*(x,y) = c(x,y,y)\alpha_{(y,x)}(T(y,x)^*)c(x,y,y)^{-1}$$

PROOF. The proof is straightforward.

□

NOTATION. Let A be the abelian subalgebra of diagonalisable operators of M, i. e.

$$A = \int_X \mathbb{C} \cdot 1_{H(x)} d\mu(x)$$

Then $A \subset Z(M)$, $A \cong L^\infty(X,\mu)$.

Proposition 5.4 $\tilde{M} \cap I(A)' = I(M)$.

PROOF. "$\supset$" is clear.
"$\subset$": Let $T \in \tilde{M} \cap I(A)'$. Let $a \in A$.
$(TI(a))(x,y) = \sum_z \alpha_{(y,z)}(T(x,z))c(x,z,y)I(a)(z,y)c(z,y,y)c(x,y,y)^{-1}$
$= T(x,y)c(x,y,y)a(y)c(x,y,y)^{-1} = T(x,y)a(y)$;
$(I(a)T)(x,y) = \sum_z \alpha_{(y,z)}(I(a)(x,z))c(x,z,y)T(z,y)c(z,y,y)c(x,y,y)^{-1}$
$= \alpha_{(y,x)}(a(x))T(x,y) = T(x,y)a(x)$.
Thus, $T(x,y)a(y) = T(y,x)a(x)$ for a. e. $(x,y) \in R$. Since this is true for every $a \in A$, the only case this can be is: $T(x,y) = 0$ for $x \neq y$ a. e. Thus, if we write $t(x) = T(x,x)$ for $x \in X$, then $t = \int_X t(x)d\mu(x) \in M$ and $T = I(t) \in I(M)$.

□

CHAPTER 6

THE EXPECTATION E

NOTATION. Let $E : \tilde{M} \to M$ be defined by the following way:

$$(ET)(x) = T(x, x), \quad T \in \tilde{M},\ x \in X$$

It is clear that E is a normal conditional expectation. By the straight calculation we get also that E is faithful.

REMARK. From now on, when speaking about an expectation from $\tilde{M}$ onto M, we identify M with $I(M)$.

CHAPTER 7

COORDINATES IN $\tilde{M}^{\nabla}$

We define coordinates in $\tilde{M}^{\nabla}$ by the similar way as it was done for $\tilde{M}$. These coordinates are analogous to the function $b(T')$ in [15].

NOTATION. For every $f \in H$, $T' \in \tilde{M}^{\nabla}$, $\tau \in \Sigma$ let

$$\begin{aligned} \tilde{f}(x,y) &= \delta_{(x,y)} f(y) \\ (T'_\tau f)(x) &= D(x,\tau^{-1}x)^{-1/2} U_{(x,\tau^{-1}x)} (T'\tilde{f})(x,\tau^{-1}x) \end{aligned}$$

Then $\tilde{f} \in \tilde{H}$, $T'_\tau f = \int_X (T'_\tau f)(x) d\mu(x)$ belongs to H. Indeed,

$$\begin{aligned} &\|T'_\tau f\|^2 = \int_X \|T'_\tau f(x)\|^2 d\mu(x) = \int_X \|(T'\tilde{f})(x,\tau^{-1}x)\|^2 D(x,\tau^{-1}x)^{-1} d\mu(x) \\ &\leq \int_X \sum_{y\sim x} \|(T'\tilde{f})(x,y)\|^2 D(x,y)^{-1} d\mu(x) \\ &= \int_R \|(T'\tilde{f})(x,y)\|^2 D(x,y)^{-1} d\nu^{-1}(x,y) \\ &= \int_R \|(T'\tilde{f})(x,y)\|^2 d\nu(x,y) = \|T'\tilde{f}\|^2 < \infty \end{aligned}$$

But $\|\tilde{f}\|_{\tilde{H}} = \|f\|_H$. Hence, it follows from the same calculation that T'_τ is bounded and $\|T'_\tau\| \leq \|T'\|$.

It is easy to see that $T'_\phi \in M'$.

NOTATION. Let for $\tau \in \Sigma$, a. e. $x \in X$:

$$T'(x,\tau^{-1}x) = D(x,\tau^{-1}x)^{1/2} \alpha_{(\tau^{-1}x,x)}(T'_\tau(x))$$

where $T'_\tau = \int_X T'_\tau(x) d\mu(x)$ is the decomposition of T'_τ.

The notation above is well defined for a. e. $x \in X$, i. e. it depends only on $\tau^{-1}x$, not on $\tau \in \Sigma$.

Proposition 7.1 *For $a' \in M'$, $\tau \in \Sigma$, a. e. $(x,y) \in R$:*

$$(I'(a'))(x,y) = \begin{cases} 0 & \text{if } x \neq y \\ a'(y) & \text{if } x = y \end{cases},$$

$$(\pi'(\tau))(x,y) = \begin{cases} 0 & \text{if } y \neq \tau x \\ D(x,y)^{1/2} \cdot 1_{H(y)} & \text{if } y = \tau x \end{cases}$$

PROOF. The proof is straightforward.

□

Theorem 7.2 (See [15, Proposition 2.6].) *For $T', T_1', T_2' \in \tilde{M}^\nabla$, $f \in H$, a. e. $(x,y) \in R$:*

$$(T'f)(x,y) = \sum_{z\sim y} c(x,z,y)T'(z,y)U_{(y,z)}c(x,z,z)^{-1}f(x,z) \tag{7.1}$$

$$(T_1'T_2')(x,y) = \sum_{z\sim y} c(x,z,y)T_1'(z,y)\alpha_{(y,z)}(c(x,z,z)^{-1}T_2'(x,z)) \tag{7.2}$$

where the series in (7.1) is understood with respect to the norm topology in $H(y)$, the series in (7.2) — with respect to the strong operator topology in $M'(y)$.

PROOF.

1. (a) For a vector $f \in \tilde{H}$ such that $\operatorname{supp} f \in \Gamma(\tau)$ with $\tau \in Sigma$, the checking is straightforward.
   (b) Let $f \in \tilde{H}_f$. The proof in this case is just a repetition of the part (1b) of the proof of Theorem 5.2.
   (c) Let f be arbitrary. Again, the proof is just a repetition of the part (1c) of the proof of Theorem 5.2.
2. Applying twice the argument of part (1) of this proof, we get:

$$(T_1'T_2'f)(x,y)$$
$$= \textstyle\sum_z c(x,z,y)T_1'(z,y)U_{(y,z)}c(x,z,z)^{-1}T_2'(x,z)U_{(z,x)}f(x,x)$$

On the other hand, by the part (1):

$$(T_1'T_2'f)(x,y)$$
$$= \textstyle\sum_w c(x,w,y)(T_1'T_2')(w,y)U_{(y,w)}c(x,w,w)^{-1}f(x,w)$$
$$= c(x,x,y)(T_1'T_2')(x,y)U_{(y,x)}c(x,x,x)^{-1}f(x,x)$$
$$= (T_1'T_2')(x,y)U_{(y,x)}f(x,x)$$

This is true for every f as above. Hence,

$$(T_1'T_2')(x,y)U_{(y,x)}$$
$$= \textstyle\sum_z c(x,z,y)T_1'(z,y)U_{(y,z)}c(x,z,z)^{-1}T_2'(x,z)U_{(x,z)}$$

a. e. , where the series is understood as the strong operator limit. Finally,

$$\begin{aligned}(T_1'T_2')(x,y) &= \sum_z c(x,z,y)T_1'(z,y)U_{(y,z)}c(x,z,z)^{-1}T_2'(x,z)U_{(z,y)} \\ &= \sum_z c(x,z,y)T_1'(z,y)\alpha_{(y,z)}(c(x,z,z)^{-1}T_2'(x,z))\end{aligned}$$

□

Theorem 7.3 *For $T' \in \tilde{M}^\nabla$, a. e. $(x,z) \in R$:*

$$T'^*(z,x) = D(z,x)c(z,x,x)\alpha_{(x,z)}(T'(x,z)^*c(x,z,z))$$

PROOF. The proof is straightforward.

□

Proposition 7.4 $\tilde{M}^\nabla \cap I'(A)' = I'(M')$.

PROOF. "$\supset$" is clear.
"$\subset$": Let $T' \in \tilde{M}^\nabla \cap I'(A)'$. Let $a \in A$.

$$(T'I'(a))(x,y) = \sum_z c(x,z,y)T'(z,y)\alpha_{(y,z)}(c(x,z,z)^{-1}I'(a)(x,z))$$
$$= T'(x,y)\alpha_{(y,x)}(a(x)) = T(x,y)a(x);$$

$$(I'(a)T')(x,y) = \sum_z c(x,z,y)I'(a)(z,y)\alpha_{(y,z)}(c(x,z,z)^{-1}T'(x,z))$$
$$= c(x,y,y)a(y)c(x,y,y)^{-1}T'(x,y) = a(y)T'(x,y) = T'(x,y)a(y)$$

Thus, $T'(x,y)a(y) = T'(x,y)a(x)$ for a. e. $(x,y) \in R$. This is true for every $a \in A$. The only case it can be is: $T'(x,y) = 0$ for $x \neq y$. Thus if we take $t'(x) = T'(x,x)$ for $x \in X$ then $t' \in M'$ and $T' = I'(t') \in I'(M')$.

$\square$

CHAPTER 8

THE EXPECTATION E'

NOTATION. Let $E' : \tilde{M}^{\nabla} \to M'$ be defined by the following:

$$(E'T')(x) = T'(x,x), \quad T' \in \tilde{M}^{\nabla}, \quad x \in X$$

It is clear that E' is a normal conditional expectation. By the straight calculation we can also get that E' is faithful.

REMARK. From now on, when speaking about an expectation from $\tilde{M}^{\nabla}$ onto M', we identify M' with $I'(M')$.

NOTATION. Let $P_{\triangle}$ be the projection acting on $\tilde{H}$ which is defined as follows:

$$(P_{\triangle} f)(x,y) = \begin{cases} 0 & \text{if } x \neq y \\ f(y,y) & \text{if } x = y \end{cases}$$

for $f \in \tilde{H}$.

Proposition 8.1 *For $f \in H$, $T \in \tilde{M}$, $T' \in \tilde{M}^{\nabla}$:*

$$\begin{aligned} (\widetilde{ET})f &= P_{\triangle} T \tilde{f} \\ (\widetilde{E'T'})f &= P_{\triangle} T' \tilde{f} \end{aligned}$$

PROOF. The proof is straightforward. □

CHAPTER 9

TOMITA-TAKESAKI THEORY FOR $\tilde{M}$ AND $\tilde{M}^{\nabla}$

Lemma 9.1 *$\tilde{\phi}_0$ is separating and cyclic for $\tilde{M}$.*

PROOF. Let $T \in \tilde{M}$, $T\tilde{\phi}_0 = 0$ and hence $T^*T\tilde{\phi}_0 = 0$.

$$(T^*T\tilde{\phi}_0)(x,x) = \sum_z \alpha_{(x,z)}((T^*T)(x,z))c(x,z,x)\tilde{\phi}_0(z,x)$$
$$= (T^*T)(x,x)\phi_0(x) = 0$$

for a. e. $x \in X$, so $E(T^*T) = 0$. But E is faithful, so $T^*T = 0$ and $T = 0$.

Thus, $\tilde{\phi}_0$ is separating for $\tilde{M}$.

Now, let $\tilde{\psi} \in \tilde{H}$, $\varepsilon > 0$, $R = \bigcup_{i=1}^{\infty} \Gamma(\tau_i)$ be as above a partition of R, where τ_i are partial Borel automorphisms of X. So, $\tilde{\psi} = \sum_{i=1}^{\infty} \chi_{\Gamma(\tau_i)}\tilde{\psi}$.

We have to approximate $\tilde{\psi}$ by $T\tilde{\phi}_0$ with $T \in \tilde{M}$. To do this it is enough to approximate every $\chi_{\Gamma(\tau_i)}\tilde{\psi}$, or every $\tilde{\psi}_i = \pi(\tau_i)\chi_{\Gamma(\tau_i)}\tilde{\psi}$. But $\operatorname{supp}\tilde{\psi}_i \subset \Delta$. Let $\psi_i \in H$, $\psi_i(x) = \tilde{\psi}_i(x,x)$ for $x \in X$. There exist $a \in M$ such that $\|a\phi_0 - \psi_i\| < \varepsilon$. Then:

$$\|I(a)\tilde{\phi}_0 - \tilde{\psi}_i\|_{\tilde{H}} = \|\widetilde{a\phi_0 - \psi_i}\|_{\tilde{H}} = \|a\phi_0 - \psi_i\|_H < \varepsilon$$

□

Now, let $\tilde{S}$, $\tilde{F}$, $\tilde{\Delta}$ and $\tilde{J}$ be the canonical operators from the Tomita-Takesaki theory associated with $\tilde{M}$ and $\tilde{\phi}_0$.

Lemma 9.2 *$\tilde{\phi}_0$ is separating and cyclic for $\tilde{M}^{\nabla}$.*

PROOF. Let $\tilde{M}'$ be the commutant of $\tilde{M}$. Then $\tilde{M}^{\nabla} \subset \tilde{M}'$. The vector $\tilde{\phi}_0$ is cyclic for $\tilde{M}$, so it is separating for $\tilde{M}'$, and in particular for $\tilde{M}^{\nabla}$.

Now, let $\tilde{\psi} \in \tilde{H}$, $\varepsilon > 0$, $R = \sum_{i=1}^{\infty} \Gamma(\tau_i)$ be the partition of R. Then $\tilde{\psi} = \sum_{i=1}^{\infty} \chi_{\Gamma(\tau_i)}\tilde{\psi}$. We have to approximate $\tilde{\psi}$ by $T'\tilde{\phi}_0$ with $T' \in \tilde{M}^{\nabla}$. To do this , it is enough to approximate every $\chi_{\Gamma(\tau_i)}\tilde{\psi}$, or every $\tilde{\psi}_i' = \pi'(\tau_i)\chi_{\Gamma(\tau_i)}\tilde{\psi}$. But $\operatorname{supp}\tilde{\psi}_i' \subset \Delta$. Let $\psi_i \in H$, $\psi_i(x) = \tilde{\psi}_i(x,x)$ for $x \in X$. There exist $a' \in M'$ such that $\|a'\phi_0 - \psi_i\|_H < \varepsilon$. Then:

$$\|I'(a')\tilde{\phi}_0 - \tilde{\psi}_i\|_{\tilde{H}} = \|\widetilde{a'\phi_0 - \psi_i}\|_{\tilde{H}} = \|a'\phi_0 - \psi_i\|_H < \varepsilon$$

□

NOTATION. For every Borel subset E of X, let P_E and Q_E be the projections in $B(\tilde{H})$ defined by the following way:

$$\begin{aligned}(P_E f)(x,y) &= \chi_E(x)f(x,y), \quad f \in \tilde{H} \\ (Q_E f)(x,y) &= \chi_E(y)f(x,y), \quad f \in \tilde{H}\end{aligned}$$

We have: P_E and Q_E are really projections, $P_E \in \tilde{M}$, $Q_E \in \tilde{M}^\nabla$ and $P_E\tilde{\phi}_0 = Q_E\tilde{\phi}_0$.

Lemma 9.3 *1. Let T be a linear or conjugate-linear bounded operator on $\tilde{H}$ such that for every Borel subset E of X: $TP_E = P_E T$, $TQ_E = Q_E T$. Then the operator T acts as follows:*

$$(Tf)(x,y) = T(x,y)f(x,y)\,, \quad f \in \tilde{H}$$

where for every $x \sim y$, $T(x,y)$ is an operator on $H(y)$.

2. Let T be a linear or conjugate-linear bounded operator on $\tilde{H}$ such that for every Borel subset E of X: $TP_E = Q_E T$, $TQ_E = P_E T$. Then the operator T acts as follows:

$$(Tf)(x,y) = T(y,x)f(y,x)\,, \quad f \in \tilde{H}$$

where for every $x \sim y$, $T(y,x)$ is an operator from $H(x)$ to $H(y)$.

PROOF. First, let T be linear.

1. If T commutes with P_E and Q_E for every Borel subset E of X, then T commutes with every $\chi_{E\times F}$ for Borel subsets E and F of X, hence T commutes with χ_G for every subset G of $X \times X$ which is an open subset of R (in the topology induced from the product topology in $X \times X$). Hence, T commutes with χ_G for every Borel subset G of R.

   By [10, Ch. II, §2, n. 5, Théorème 1] T is a decomposable operator on $\tilde{H} = \int_R H(x,y)d\nu(x,y)$, i. e. $(Tf)(x,y) = T(x,y)f(x,y)$.

2. Let $\tilde{\theta} \in B(\tilde{H})$ be defined as follows:

$$(\tilde{\theta}f)(x,y) = D(x,y)^{1/2}U_{(y,x)}f(y,x)$$

   Then:

   (a) $\tilde{\theta}$ is unitary;

   (b) $\tilde{\theta}^{-1} = \tilde{\theta}$;

   (c) $\tilde{\theta}P_E\tilde{\theta} = Q_E$, $\tilde{\theta}Q_E\tilde{\theta} = P_E$.

   Let $\bar{T} = T\tilde{\theta}$. Then:

$$\bar{T}P_E = T\tilde{\theta}P_E = T\tilde{\theta}P_E\tilde{\theta}\tilde{\theta} = TQ_E\tilde{\theta} = P_E T\tilde{\theta} = P_E\bar{T}$$
$$\bar{T}Q_E = T\tilde{\theta}Q_E = T\tilde{\theta}Q_E\tilde{\theta}\tilde{\theta} = TP_E\tilde{\theta} = Q_E T\tilde{\theta} = Q_E\bar{T}$$

By part (1) of this proof: $(\bar{T}f)(x,y) = \bar{T}(x,y)f(x,y)$, $f \in \tilde{H}$. Then:

$$(Tf)(x,y) = (\bar{T}\tilde{\theta}f)(x,y) = \bar{T}(x,y)(\tilde{\theta}f)(x,y)$$
$$= \bar{T}(x,y)D(x,y)^{1/2}U_{(y,x)}f(y,x)$$

Now, let T be conjugate-linear. Then T is linear as an operator from $\tilde{H}$ to $\tilde{H}'$ where $\tilde{H}'$ is the dual space to $\tilde{H}$ (i. e. the same space with the conjugate inner product). All the considerations are correct in this case too.

□

REMARK. The same fact is true for a non-bounded self-ajoint operator, which is a strong limit of bounded operators.

Let us study some structural properties of the operators $\tilde{S}$, $\tilde{F}$, $\tilde{\Delta}$ and $\tilde{J}$. In what follows, $\mathcal{D}$ denotes the donains of non-bounded operators.

Lemma 9.4 *For a Borel subset E of X:*

$$Q_E\tilde{S} \subset \tilde{S}P_E\,, \quad P_E\tilde{S} \subset \tilde{S}Q_E\,, \quad P_E\tilde{F} \subset \tilde{F}Q_E\,, \quad Q_E\tilde{F} \subset \tilde{F}P_E$$

PROOF.

1. The straight calculation shows that for $f = T\tilde{\phi}_0$ with $T \in \tilde{M}$:

$$\tilde{S}P_E f = Q_E\tilde{S}f$$

Now, if $f \in \mathcal{D}(\tilde{S}) = \mathcal{D}(\tilde{\Delta}^{1/2})$ then there exists a sequence $(f_n) \subset \tilde{H}$ such that $f_n = T_n\tilde{\phi}_0$, $T_n \in \tilde{M}$, $f_n \to f$ and $\tilde{S}f_n \to \tilde{S}f$. We have: $\tilde{S}P_E f_n = Q_E\tilde{S}f_n$ for every n. Moreover, $Q_E\tilde{S}f_n \to Q_E\tilde{S}f$. Thus, $P_E f_n \in \mathcal{D}(\tilde{S})$, $P_E f_n \to P_E f$ and $\tilde{S}P_E f_n \to Q_E\tilde{S}f$. But $\tilde{S}$ is closed, hence $P_E f \in \mathcal{D}(\tilde{S})$ and $\tilde{S}P_E f = Q_E\tilde{S}f$ for $f \in \mathcal{D}(\tilde{S})$.

   Thus, $Q_E\tilde{S} \subset \tilde{S}P_E$.

2. Now, $\tilde{S}Q_E T\tilde{\phi}_0 = \tilde{S}TQ_E\tilde{\phi}_0 = \tilde{S}TP_E\tilde{\phi}_0 = (TP_E)^*\tilde{\phi}_0 = P_E T^*\tilde{\phi}_0$ $= P_E\tilde{S}T\tilde{\phi}_0$. Thus, for $f = T\tilde{\phi}_0$:

$$\tilde{S}Q_E f = P_E\tilde{S}f$$

   If $f \in \mathcal{D}(\tilde{S}) = \mathcal{D}(\tilde{\Delta}^{1/2})$ then there exists a sequence $(f_n) \subset \tilde{H}$ such that $f_n = T_n\tilde{\phi}_0$, $T_n \in \tilde{M}$, $f_n \to f$ and $\tilde{S}f_n \to \tilde{S}f$. We have: $\tilde{S}Q_E f_n = P_E\tilde{S}f_n$ for every n. Moreover, $P_E\tilde{S}f_n \to P_E\tilde{S}f$. Thus, $Q_E f_n \in \mathcal{D}(\tilde{S})$, $Q_E f_n \to Q_E f$ and $\tilde{S}Q_E f_n \to P_E\tilde{S}f$. But $\tilde{S}$ is closed, hence $Q_E f \in \mathcal{D}(\tilde{S})$ and $\tilde{S}Q_E f = P_E\tilde{S}f$.

   Thus, $P_E\tilde{S} \subset \tilde{S}Q_E$.

3. It follows from (1) that

$$P_E\tilde{F} \subset \overline{P_E\tilde{F}} = (P_E\tilde{F})^{**} = (\tilde{S}P_E)^* \subset (Q_E\tilde{S})^* = \tilde{F}Q_E$$

4. It follows from (2) that

$$Q_E\tilde{F} \subset \overline{Q_E\tilde{F}} = (Q_E\tilde{F})^{**} = (\tilde{S}Q_E)^* \subset (P_E\tilde{S})^* = \tilde{F}P_E$$

□

Lemma 9.5 $P_E\tilde{\Delta} \subset \tilde{\Delta}P_E$, $Q_E\tilde{\Delta} \subset \tilde{\Delta}Q_E$.

PROOF.

$$P_E\tilde{\Delta} = \overline{P_E\tilde{F}\tilde{S}} \subset \overline{\tilde{F}Q_E\tilde{S}} \subset \overline{\tilde{F}\tilde{S}P_E} = \overline{\tilde{\Delta}P_E} = \tilde{\Delta}P_E$$

$$Q_E\tilde{\Delta} = \overline{Q_E\tilde{F}\tilde{S}} \subset \overline{\tilde{F}P_E\tilde{S}} \subset \overline{\tilde{F}\tilde{S}Q_E} = \overline{\tilde{\Delta}Q_E} = \tilde{\Delta}Q_E$$

□

Corollary 9.6 *For every $(x, y) \in R$ there exists an operator $\tilde{\Delta}(x, y)$ on $H(y)$ such that for every $f \in \mathcal{D}(\tilde{\Delta})$, for a. e. $(x, y) \in R$:*

$$(\tilde{\Delta}f)(x, y) = \tilde{\Delta}(x, y)f(x, y)$$

Corollary 9.7 *1. $P_E\tilde{\Delta}^{1/2} \subset \tilde{\Delta}^{1/2}P_E$, $\quad Q_E\tilde{\Delta}^{1/2} \subset \tilde{\Delta}^{1/2}Q_E$;*

2. $P_E\tilde{\Delta}^{-1/2} \subset \tilde{\Delta}^{-1/2}P_E$, $\quad Q_E\tilde{\Delta}^{-1/2} \subset \tilde{\Delta}^{-1/2}Q_E$;

3. $P_E\tilde{\Delta}^{-1} \subset \tilde{\Delta}^{-1}P_E$, $\quad Q_E\tilde{\Delta}^{-1} \subset \tilde{\Delta}^{-1}Q_E$.

PROOF. This follows from Lemma 9.5 above and the spectral theorem for non-bounded self-ajoint operators.

□

Corollary 9.8 *For $(x, y) \in R$ there exist operators $\tilde{\Delta}^{1/2}(x, y)$, $\tilde{\Delta}^{-1/2}(x, y)$, $\tilde{\Delta}^{-1}(x, y)$ such that for a. e. $(x, y) \in R$:*

$$\begin{array}{lcll} (\tilde{\Delta}^{1/2}f)(x, y) & = & \tilde{\Delta}^{1/2}(x, y)f(x, y), & f \in \mathcal{D}(\tilde{\Delta}^{1/2}) \\ (\tilde{\Delta}^{-1/2}f)(x, y) & = & \tilde{\Delta}^{-1/2}(x, y)f(x, y), & f \in \mathcal{D}(\tilde{\Delta}^{-1/2}) \\ (\tilde{\Delta}^{-1}f)(x, y) & = & \tilde{\Delta}^{-1}(x, y)f(x, y), & f \in \mathcal{D}(\tilde{\Delta}^{-1}) \end{array}$$

Lemma 9.9 $\tilde{J}P_E = Q_E\tilde{J}$, $\tilde{J}Q_E = P_E\tilde{J}$.

PROOF.

1. $\tilde{S} = \tilde{J}\tilde{\Delta}^{1/2}$, hence $\tilde{S}\tilde{\Delta}^{-1/2} = \tilde{J}\tilde{\Delta}^{1/2}\tilde{\Delta}^{-1/2} \subset \tilde{J}$ and $\mathcal{D}(\tilde{S}\tilde{\Delta}^{-1/2}) = \mathcal{D}(\tilde{\Delta}^{-1/2})$. So, $Q_E\tilde{S}\tilde{\Delta}^{-1/2} \subset Q_E\tilde{J}$.

   On the other hand:

   $$Q_E\tilde{S}\tilde{\Delta}^{-1/2} \subset \tilde{S}P_E\tilde{\Delta}^{-1/2} \subset \tilde{S}\tilde{\Delta}^{-1/2}P_E \subset \tilde{J}P_E$$

   But $Q_E\tilde{S}\tilde{\Delta}^{-1/2}$ is densely defined, and $Q_E\tilde{J}$ and $\tilde{J}P_E$ are bounded. Hence:

   $$Q_E\tilde{J} = \tilde{J}P_E$$

2. $\tilde{F} = \tilde{J}\tilde{\Delta}^{-1/2}$, hence $\tilde{F}\tilde{\Delta}^{1/2} = \tilde{J}\tilde{\Delta}^{-1/2}\tilde{\Delta}^{1/2} \subset \tilde{J}$ and $\mathcal{D}(\tilde{F}\tilde{\Delta}^{1/2}) = \mathcal{D}(\tilde{\Delta}^{1/2})$. So, $P_E\tilde{F}\tilde{\Delta}^{1/2} \subset P_E\tilde{J}$.

   On the other hand:

   $$P_E\tilde{F}\tilde{\Delta}^{1/2} \subset \tilde{F}Q_E\tilde{\Delta}^{1/2} \subset \tilde{F}\tilde{\Delta}^{1/2}Q_E \subset \tilde{J}Q_E$$

   But $P_E\tilde{F}\tilde{\Delta}^{1/2}$ is densely defined, and $P_E\tilde{J}$ and $\tilde{J}$ are bounded. Hence:

   $$P_E\tilde{J} = \tilde{J}Q_E$$

□

Corollary 9.10 *For every $(y,x) \in R$ there exists an operator $\tilde{J}(y,x)$ acting from $H(x)$ to $H(y)$ such that for a. e. $(x,y) \in R$:*

$$(\tilde{J}f)(x,y) = \tilde{J}(y,x)f(y,x)\ , \quad f \in \tilde{H}$$

Corollary 9.11 *For every $(y,x) \in R$ there exists operators $\tilde{S}(y,x)$, $\tilde{F}(y,x)$ acting from $H(x)$ to $H(y)$ such that for a. e. $(x,y) \in R$:*

$$\begin{aligned}(\tilde{S}f)(x,y) &= \tilde{S}(y,x)f(y,x)\ , \quad f \in \mathcal{D}(\tilde{S})\\ (\tilde{F}f)(x,y) &= \tilde{F}(y,x)f(y,x)\ , \quad f \in \mathcal{D}(\tilde{F})\end{aligned}$$

PROOF.

1. For $f \in \mathcal{D}(\tilde{S}) = \mathcal{D}(\tilde{\Delta}^{1/2})$:

$$\begin{aligned}&(\tilde{S}f)(x,y) = (\tilde{J}\tilde{\Delta}^{1/2})(x,y) = \tilde{J}(y,x)(\tilde{\Delta}^{1/2}f)(y,x)\\ &= \tilde{J}(y,x)\tilde{\Delta}^{1/2}(y,x)f(y,x)\end{aligned}$$

and it is enough to take $\tilde{S}(y,x) = \tilde{J}(y,x)\tilde{\Delta}^{1/2}(y,x)$.

2. For $f \in \mathcal{D}(\tilde{F}) = \mathcal{D}(\tilde{\Delta}^{-1/2})$:

$$\begin{aligned}&(\tilde{F}f)(x,y) = (\tilde{J}\tilde{\Delta}^{-1/2}f)(y,x) = \tilde{J}(y,x)(\tilde{\Delta}^{-1/2}f)(y,x)\\ &= \tilde{J}(y,x)\tilde{\Delta}^{-1/2}(y,x)f(y,x)\end{aligned}$$

and it is enough to take $\tilde{F}(y,x) = \tilde{J}(y,x)\tilde{\Delta}^{-1/2}(y,x)$.

□

Corollary 9.12 $\tilde{J}(x,y) = \tilde{J}(y,x)^{-1}$ *a. e.*

PROOF. $\tilde{J}^2 = 1_{\tilde{H}}$ so for $f \in \tilde{H}$:

$$\tilde{f}(x,y) = \tilde{J}(\tilde{J}f)(x,y) = \tilde{J}(y,x)(\tilde{J}f)(y,x) = \tilde{J}(y,x)\tilde{J}(x,y)f(x,y)$$

□

Corollary 9.13 $\tilde{S}(x,y) = \tilde{S}(y,x)^{-1}$, $\tilde{F}(x,y) = \tilde{F}(y,x)^{-1}$ *a. e.*

The proof is similar to the proof of Corollary 9.12 and is omitted.

Theorem 9.14 *$\tilde{M}^{\nabla}$ is the commutant of $\tilde{M}$.*

PROOF. Let $\tilde{M}'$ be the commutant of $\tilde{M}$. The inclusion $\tilde{M}^{\nabla} \subset \tilde{M}'$ was already proved. It remains to prove the inclusion $\tilde{M}' \subset \tilde{M}^{\nabla}$, or $\tilde{J}\tilde{M}\tilde{J} \subset \tilde{M}^{\nabla}$.

It is enough to prove that

$$\begin{aligned}\tilde{J}I(a)\tilde{J} &\in \tilde{M}^{\nabla}, \quad a \in M\\ \tilde{J}\pi(\tau)\tilde{J} &\in \tilde{M}^{\nabla}, \quad \tau \in \Sigma\end{aligned}$$

For this, it is enough to prove that $\tilde{J}I(a)\tilde{J}\tilde{\phi}_0 \in \tilde{M}^{\nabla}\tilde{\phi}_0$ and $\tilde{J}\pi(\tau)\tilde{J}\tilde{\phi}_0 \in \tilde{M}^{\nabla}\tilde{\phi}_0$. The last two inclusions are verified immediately.

□

Corollary 9.15 $\tilde{J}I(M)\tilde{J} = I'(M')$.

PROOF. The checking is immediate. □

Now, let S, F, J, $\triangle$ be the canonical operators from the Tomita-Takesaki theory associated with M and ϕ_0. These operators commute with central projections of M, so they are decomposable: $J = \int_X J(x)d\mu(x)$, $\triangle = \int_X \triangle(x)d\mu(x)$ by [10, Ch. II, §2, n. 5, Théorème 1], hence $S = \int_X S(x)d\mu(x)$ and $F = \int_X F(x)d\mu(x)$ too.

Proposition 9.16

$$\begin{aligned}
\tilde{\triangle}^{1/2}(x,y) &= \tilde{\triangle}(x,y)^{1/2}; \\
\tilde{\triangle}^{-1/2}(x,y) &= \tilde{\triangle}^{-1}(x,y)^{1/2} = \tilde{\triangle}(x,y)^{-1/2}; \\
\triangle^{1/2}(x) &= \triangle(x)^{1/2}; \\
\triangle^{-1/2}(x) &= \triangle^{-1}(x)^{1/2} = \triangle(x)^{-1/2}
\end{aligned}$$

PROOF. This follows from the spectral theorem for self-ajoint operators. Indeed, for every polinomial p: $p(\tilde{\triangle})(x,y) = p(\tilde{\triangle}(x,y))$, $p(\triangle)(x) = p(\triangle(x))$ a. e., hence this holds for all Borel functions. □

Proposition 9.17 *For $f \in H$:*

1. *$f \in \mathcal{D}(S)$ if and only if $\tilde{f} \in \mathcal{D}(\tilde{S})$, and under these conditions:*
$$\widetilde{Sf} = \tilde{S}\tilde{f}$$
2. *$f \in \mathcal{D}(F)$ if and only if $\tilde{f} \in \mathcal{D}(\tilde{F})$, and under these conditions:*
$$\widetilde{Ff} = \tilde{F}\tilde{f}$$
3. *$f \in \mathcal{D}(\triangle)$ if and only if $\tilde{f} \in \mathcal{D}(\tilde{\triangle})$, and under these conditions:*
$$\widetilde{\triangle f} = \tilde{\triangle}\tilde{f}$$

PROOF. The proof is straightforward. □

Corollary 9.18 *For a. e. $x \in X$:*

$$\begin{aligned}
\tilde{S}(x,x) &= S(x) \\
\tilde{F}(x,x) &= F(x) \\
\tilde{\triangle}(x,x) &= \triangle(x)
\end{aligned}$$

Corollary 9.19

$$\begin{aligned}
\tilde{\triangle}^{1/2}(x,x) &= \triangle^{1/2}(x) \\
\tilde{\triangle}^{-1/2}(x,x) &= \triangle^{-1/2}(x)
\end{aligned}$$

PROOF. This follows from Proposition 9.16. □

Corollary 9.20 *1. $f \in \mathcal{D}(\triangle^{1/2})$ if and only if $\tilde{f} \in \mathcal{D}(\tilde{\triangle}^{1/2})$, and under these conditions: $\widetilde{\triangle^{1/2} f} = \tilde{\triangle}^{1/2}\tilde{f}$.*

2. $f \in \mathcal{D}(\triangle^{-1/2})$ if and only if $\tilde{f} \in \mathcal{D}(\tilde{\triangle}^{-1/2})$, and under these conditions: $\widetilde{\triangle^{-1/2} f} = \tilde{\triangle}^{-1/2}\tilde{f}$.

PROOF.

1.

$$f \in \mathcal{D}(\triangle^{1/2}) \iff \int_X \|\triangle^{1/2}(x)f(x)\|^2 d\mu(x) < \infty$$
$$\iff \int_X \|\tilde{\triangle}^{1/2}(x,x)\tilde{f}(x,x)\|^2 d\mu(x) < \infty \iff \tilde{f} \in \mathcal{D}(\tilde{\triangle}^{1/2});$$
$$(\widetilde{\triangle^{1/2} f})(x,y) = \begin{cases} (\triangle^{1/2}f)(y) & \text{if } x = y \\ 0 & \text{if } x \neq y \end{cases} = \begin{cases} \triangle^{1/2}(y)f(y) & \text{if } x = y \\ 0 & \text{if } x \neq y \end{cases}$$
$$= \begin{cases} \tilde{\triangle}^{1/2}(y,y)\tilde{f}(y,y) & \text{if } x = y \\ 0 & \text{if } x \neq y \end{cases} = (\tilde{\triangle}^{1/2}\tilde{f})(x,y)$$

2. The proof of this part is just the repetition of the part 1.

□

Proposition 9.21 *For a. e. $x \in X$: $\tilde{J}(x,x) = J(x)$.*

PROOF. $\tilde{J}(x,x)$ and $J(x)$ are bounded. For $f \in \mathcal{D}(\triangle^{-1/2})$ we have:

$$\tilde{J}(x,x)f(x) = \tilde{J}(x,x)\tilde{f}(x,x) = (\tilde{J}\tilde{f})(x,x)$$
$$= (\tilde{S}\tilde{\triangle}^{-1/2}\tilde{f})(x,x) = \tilde{S}(x,x)(\tilde{\triangle}^{-1/2}\tilde{f})(x,x)$$
$$= \tilde{S}(x,x)\tilde{\triangle}^{-1/2}(x,x)\tilde{f}(x,x) = S(x)\triangle^{-1/2}(x)f(x) = J(x)f(x)$$

But $\mathcal{D}(\triangle^{-1/2})$ is dense in H, so for a. e. $x \in X$ the set $\{f(x) \mid f \in \mathcal{D}(\triangle^{-1/2})\}$ is dense in $H(x)$. Thus $\tilde{J}(x,x)$ and $J(x)$ are equal on a dense subspace of $H(x)$, hence they are equal.

□

Proposition 9.22 *For $a \in Z(M)$:*

$$I(a)\tilde{S} \subset \tilde{S}I'(a), \qquad I'(a)\tilde{S} \subset \tilde{S}I(a)$$
$$I(a)\tilde{F} \subset \tilde{F}I'(a), \qquad I'(a)\tilde{F} \subset \tilde{F}I(a)$$

PROOF. The proof is just a repetition of the proof of Lemma 9.4, where $I(a)$ stands for P_E and $I'(a)$ for Q_E.

□

Proposition 9.23 *1. The operator $\tilde{\triangle}$ commutes with every element of $I(Z(M)) \cup I'(Z(M))$.*

2. For every $a \in Z(M)$: $\tilde{J}I(a) = I'(a)\tilde{J}$, $\tilde{J}I'(a) = I(a)\tilde{J}$.

PROOF. The proof is again a repetition of the proofs of Lemmas 9.5 and 9.9, where for $a \in Z(M)$, $I(a)$ stands for P_E, and $I'(a)$ stands for Q_E.

□

CHAPTER 10

$I(M)$-AUTOMORPHISMS OF $\tilde{M}$

NOTATION. Let $\tilde{Z} = \int_R Z(M(y))d\nu(x,y)$, $\mathcal{V} = \{V \in \tilde{Z} \mid V$ is unitary, and for a. e. $x \in X$, $y \sim z \sim x : V(z,y) = \alpha_{(y,x)}(V(z,x))c(z,x,y)V(x,y)c(z,x,y)^{-1}\}$.

For a unitary operator $U \in B(\tilde{H})$ let as usual AdU be the automorphism of $B(\tilde{H})$ defined by: Ad$U(T) = UTU^{-1}$ for $T \in B(\tilde{H})$.

Lemma 10.1 *For $V \in \mathcal{V}$:*

1. *$V(x,x) = 1_{H(x)}$ for a. e. $x \in X$;*

2. *$V(y,x) = \alpha_{(x,y)}(c(x,y,y)^{-1}V(x,y)^{-1}c(x,y,y))$ for a. e. $(x,y) \in R$.*

PROOF.

1. Take $x = y = z$:
$$V(x,x) = V(x,x) \cdot V(x,x)$$
and $V(x,x)$ is unitary a. e. By the spectral theorem we can assume that $V(x,x)$ is an operator of multiplication on some function f with values in $\mathbb{T} = \{z \in \mathbb{C} \mid |z| = 1\}$ on a space $L^2(\Omega)$ for some Ω. This function satisfies $f(\omega)^2 = f(\omega)$ a. e., so $f(\omega) = 1$ a. e., hence $V(x,x) = 1_{H(x)}$ a. e.

2. Take $z = y$:
$$1_{H(y)} = V(y,y) = \alpha_{(y,x)}(V(y,x))c(y,x,y)V(x,y)c(y,x,y)^{-1}$$
Thus,
$$V(y,x) = \alpha_{(x,y)}(c(x,y,y)^{-1}V(x,y)^{-1}c(x,y,y))$$

□

Proposition 10.2 *For every $V \in \mathcal{V}$, Ad$V|\tilde{M}$ is an automorphism of $\tilde{M}$ leaving $I(M)$ pointwise fixed.*

PROOF. The proof is straightforward.

□

Theorem 10.3 *Every automorphism of $\tilde{M}$ leaving $I(M)$ pointwise fixed has the form* $\mathrm{Ad}V|\tilde{M}$ *for some* $V \in \mathcal{V}$.

PROOF. Let λ be such an automorphism of $\tilde{M}$. It is spatial. Let it be defined by the unitary operator V, commuting with $\tilde{J}$ (i. e. let $\lambda = \mathrm{Ad}V|\tilde{M}$).

The automorphism λ leaves $I(a)$ fixed for every $a \in M$, so $VI(a)V^{-1} = I(a)$, i. e. V commutes with $I(a)$ for $a \in M$. In particular, $VP_E = P_E V$, $E \subset X$. But $\tilde{J}P_E = Q_E\tilde{J}$, hence $\tilde{J}P_E\tilde{J} = Q_E$, hence $VQ_E = V\tilde{J}P_E\tilde{J} = \tilde{J}P_E\tilde{J}V = Q_E V$. By Lemma 9.3, V acts as follows:

$$(Vf)(x,y) = V(x,y)f(x,y), \quad f \in \tilde{H}$$

where for a. e. $(x,y) \in R$, $V(x,y)$ are unitaries acting on $H(y)$. For $a \in M$:

$$\begin{aligned}&\alpha_{(y,x)}(a(x))V(x,y)f(x,y) = (I(a)Vf)(x,y)\\ &= (VI(a)f)(x,y) = V(x,y)\alpha_{(y,x)}(a(x))f(x,y)\end{aligned}$$

Hence $V(x,y) \in M'(y)$ for a. e. $(x,y) \in R$.

The operator $V^* = V^{-1}$ acts as follows:

$$(V^*f)(x,y) = V(x,y)^{-1}f(x,y), \quad f \in \tilde{H}$$

We have:

$$(\tilde{J}Vf)(x,y) = \tilde{J}(y,x)(Vf)(y,x) = \tilde{J}(y,x)V(y,x)f(y,x)$$

On the other hand,

$$\begin{aligned}&(\tilde{J}Vf)(x,y) = (V\tilde{J}f)(x,y)\\ &= V(x,y)(\tilde{J}f)(x,y) = V(x,y)\tilde{J}(y,x)f(y,x)\end{aligned}$$

Thus, $\tilde{J}(y,x)V(y,x) = V(x,y)\tilde{J}(y,x)$. In particular, $\tilde{J}(x,x)V(x,x) = V(x,x)\tilde{J}(x,x)$. But $\tilde{J}(x,x) = J(x)$, so

$$\begin{aligned}&J(x)V(x,x) = V(x,x)J(x);\\ &V(x,x) = J(x)V(x,x)J(x)\end{aligned}$$

Let $\bar{V}(x) = V(x,x)$, $\bar{V} = \int_X \bar{V}(x)d\mu(x)$. Then $\bar{V} \in M'$ and $\bar{V} = J\bar{V}J$. This follows that for a. e. $x \in X$:

1. $\bar{V} \in Z(M)$;
2. $\bar{V} = \bar{V}^*$.

Returning to $V(x,x)$ we have:

1. $V(x,x) \in Z(M(x))$ for a. e. $x \in X$;
2. $V(x,x) = V(x,x)^*$ for a. e. $x \in X$.

Let $T \in \tilde{M}$, $S = VTV^* \in \tilde{M}$, $f \in \tilde{H}$.

$$\begin{aligned}(Sf)(x,y) &= (VTV^*)(x,y) \\ &= \textstyle\sum_z V(x,y)\alpha_{(y,z)}(T(x,z))c(x,z,y)V(z,y)^* f(z,y)\end{aligned}$$

On the other hand:

$$(Sf)(x,y) = \sum_z \alpha_{(y,z)}(S(x,z))c(x,z,y)f(z,y)$$

It follows that for a. e. $(x,y) \in R$, $z \sim y$:

$$V(x,y)\alpha_{(y,z)}(T(x,z))c(x,z,y)V(z,y)^* = \alpha_{(y,z)}(S(x,z))c(x,z,y)$$

Now, let x, z, T be such that $T(x,z)$ is invertible. Then we have:

$$V(x,y)c(x,y,z)V(z,y)^* c(x,z,y)^{-1} = \alpha_{(y,z)}(T(x,z)^{-1}S(x,z))$$

Let

$$F(x,z) = T(x,z)^{-1}S(x,z) = \alpha_{(z,y)}(V(x,y)c(x,z,y)V(z,y)^* c(x,z,y)^{-1})$$

Then $F(x,z)$ does not depend on T, S and y, and $F(x,z) \in M(z)$ a. e. Let $y = x$. Then

$$F(x,z) = \alpha_{(z,x)}(V(x,x)c(x,z,x)V(z,x)^* c(x,z,x)^{-1})$$

So, $V(x,x)c(x,z,x)V(z,x)^* c(x,z,x)^{-1} \in M(x)$. But $V(x,x) \in Z(M(x))$, so $c(x,z,x)V(z,x)^* c(x,z,x)^{-1} \in M(x)$ hence $V(z,x)^* \in M(x)$ and $V(z,x) \in M(x)$.

It was proved above that $V(x,y) \in M'(y)$ a. e. Thus, $V(x,y) \in Z(M(y))$ a. e. and $V \in \tilde{Z}$. Moreover, $F(x,z)$ does not depend on y, so:

$$\begin{aligned}F(x,z) &= \alpha_{(z,x)}(V(x,x)c(x,z,x)V(z,x)^* c(x,z,x)^{-1}) \\ &= \alpha_{(z,y)}(V(x,y)c(x,z,y)V(z,y)^* c(x,z,y)^{-1})\end{aligned}$$

Hence:

$$\begin{aligned}V(x,x) \quad = \quad & \alpha_{(x,y)}(V(x,y)c(x,z,y)V(z,y)^* c(x,z,y)^{-1} \\ & \qquad\qquad c(x,z,x)V(z,x)c(x,z,x)^{-1}\end{aligned} \tag{10.1}$$

b Let $W \in B(\tilde{H})$ be the operator acting as follows:

$$(Wf)(x,y) = c(x,y,y)V(y,y)c(x,y,y)^{-1}f(x,y), \quad f \in \tilde{H}$$

Then W is unitary, $W \in \tilde{Z}$ and $W^* = W$. The straight calculation shows that $WTW^* = T$, i. e. $W \in \tilde{M}'$. Let $V' = VW$. Then $V' \in \tilde{Z}$, $\mathrm{Ad}V'|\tilde{M} = \mathrm{Ad}V|\tilde{M}$ and $(V'f)(x,y) = V'(x,y)f(x,y)$ with $V'(x,y) = V(x,y)W(x,y)$. The straight calculation shows that for $x \sim y \sim z$:

$$V'(z,y) = \alpha_{(y,z)}(V'(z,x))c(x,z,y)^{-1}V'(z,y)c(x,z,y)$$

Thus, $V' \in \mathcal{V}$, and $\lambda = \mathrm{Ad}V'$.

□

Theorem 10.4 *If* $V_1, V_2 \in \mathcal{V}$ *and* $\mathrm{Ad}V_1|\tilde{M} = \mathrm{Ad}V_2|\tilde{M}$, *then* $V_1 = V_2$.

PROOF. $\mathcal{V}$ is a group, so it is enough to prove the following: if $V \in \mathcal{V}$ and $\mathrm{Ad}V|\tilde{M} = \mathrm{id}_{\tilde{M}}$ then $V = 1_{\tilde{H}}$.

Let $V \in \mathcal{V}$ and $\mathrm{Ad}V|\tilde{M} = \mathrm{id}_{\tilde{M}}$ i. e. $V \in \tilde{M}'$. The operator V commutes with $I'(A)$ so $V \in \tilde{M}' \cap I'(A)' = I'(M')$. Let $V = I'(\bar{V})$ with $\bar{V} \in M'$. Then:

$$\begin{aligned} V(x,y)f(x,y) = (Vf)(x,y) &= (I'(\bar{V})f)(x,y) \\ &= c(x,y,y)\bar{V}(y)c(x,y,y)^{-1}f(x,y), \quad f \in \tilde{H} \end{aligned}$$

Thus, for a. e. $(x,y) \in R$: $V(x,y) = c(x,y,y)\bar{V}(y)c(x,y,y)^{-1}$ and $\bar{V}(y) = c(x,y,y)^{-1}V(x,y)c(x,y,y)$. Take $y = x$:

$$\bar{V}(y) = c(y,y,y)^{-1}V(y,y)c(y,y,y) = V(y,y) = 1_{H(y)}$$

Hence $V(x,y) = 1_{H(y)}$, i. e. $V = 1_{\tilde{H}}$.

□

CHAPTER 11

FLOWS OF AUTOMORPHISMS

NOTATION. Let $M(x,y) \equiv M(y)$, $(x,y) \in R$.

Let for every $(x,y) \in R$:

$$\begin{aligned} Z(M(x,y)) &\cong L^\infty(Y_{(x,y)}, \mathcal{M}_{(x,y)}, m_{(x,y)}) \\ H(x,y) &\cong L^2(Y_{(x,y)}, \mathcal{M}_{(x,y)}, m_{(x,y)}) \end{aligned}$$

where all $Y_{(x,y)}$ are pairwise disjoint sets. We will identify $Z(M(x,y))$ with $L^\infty(Y_{(x,y)})$ and $H(x,y)$ with $L^2(Y_{(x,y)})$.

Let $Y = \bigcup_{(x,y)\in R} Y_{(x,y)}$. Let $\mathcal{M}$ be the following σ-algebra:

$\mathcal{M} = \{A \subset Y \mid$ for a. e. $(x,y) \in R$, $A \cap Y_{(x,y)} \in \mathcal{M}_{(x,y)}$, and the function $(x,y) \mapsto m_{(x,y)}(A \cap Y_{(x,y)}$ is measurable $\}$.

Let m be the following measure on Y:

$$m(A) = \int_R m_{(x,y)}(A \cap Y_{(x,y)}) d\nu(x,y), \quad A \in \mathcal{M}$$

For $x \sim y \sim z$ we consider the isomorphisms $\alpha_{(y,x)} : M(x) \to M(y)$ and $\mathrm{Ad}(c(z,x,y)) : M(y) \to M(y)$ as isomorphisms from $Z(M(z,x))$ onto $Z(M(z,y))$ and from $Z(M(x,y))$ onto $Z(M(z,y))$ respectively.

NOTATION. Let $\bar{\alpha}^z_{(y,x)}$ be the Borel isomorphism from $Y_{(z,y)}$ to $Y_{(z,x)}$ defined by

$$(\alpha_{(y,x)}(f))(s) = f(\bar{\alpha}^z_{(y,x)}(s)), \qquad s \in Y_{(z,y)}, \quad f \in Z(M(z,x))$$

Let $\bar{c}(z,x,y)$ be the Borel isomorphism from $Y_{(z,y)}$ to $Y_{(x,y)}$ defined by

$$(c(z,x,y) f c(z,x,y)^{-1})(s) = f(\bar{c}(z,x,y)(s)), \qquad s \in Y_{(z,y)}, \quad f \in Z(M(x,y))$$

For any measure space C and a topological space D let $U(C,D)$ be the set of all classes of measurable maps from C to D.

For every map $f : Y \to D$ and for every $(x,y) \in R$, let $f(x,y)$ be the restriction of f to $Y_{(x,y)}$, i. e. $f(x,y) : Y_{(x,y)} \to D$.

For an abelian locally compact group Γ let $Z^1(Y,\Gamma)$ be the set of all $f \in U(Y,\Gamma)$ which satisfy:

$$f(z,y)(s) = f(z,x)(\bar{\alpha}^z_{(y,x)}(s)) \cdot f(x,y)(\bar{c}(z,x,y)(s))$$

for a. e. $x \sim y \sim z$, $s \in Y_{(z,y)}$.

$Z^1(Y, \Gamma)$ is a group with respect to the pointwise multiplication. In particular, $Z^1(Y, \mathbb{T}) \cong \mathcal{V}$.

In what follows, let G be any abelian locally compact second countable group with the dual group $\hat{G}$.

NOTATION. For every $\gamma \in Z^1(Y, \hat{G})$ and every $g \in G \cong \hat{\hat{G}}$ we define an element $g(\gamma) \in U(Y, \mathbb{T})$ as follows: $g(\gamma)(s) = g(\gamma(s))$, where $s \in Y$.

The straight calculation shows that $g(\gamma) \in Z^1(Y, \mathbb{T})$.

Theorem 11.1 *For every $\gamma \in Z^1(Y, \hat{G})$ the map $\phi_\gamma : g \mapsto g(\gamma)$ is a continuous homomorphism from G to $Z^1(Y, \mathbb{T})$, where the topology on $Z^1(Y, \mathbb{T})$ is the topology of convergence in some finite measure equivalent to the measure m on Y.*

Conversely, every continuous homomorphism from G to $Z^1(Y, \mathbb{T})$ is of the form ϕ_γ for a unique $\gamma \in Z^1(Y, \hat{G})$.

PROOF.

1. For $g_1, g_2 \in G$, $s \in Y$:

$$\begin{aligned}\phi_\gamma(g_1g_2)(s) &= (g_1g_2)(\gamma)(s) = (g_1g_2)(\gamma(s)) \\ &= g_1(\gamma(s))g_2(\gamma(s)) = g_1(\gamma)(s)g_2(\gamma)(s) = \phi_\gamma(g_1)(s)\phi_\gamma(g_2)(s)\end{aligned}$$

Thus, ϕ_γ is a homomorphism.

2. Let $g_n \to g$ in G. Then $g_n \to g$ pointwise on $\hat{G}$ so for $s \in Y$, $g_n(\gamma(s)) \to g(\gamma(s))$ hence $g_n(\gamma)(s) \to g(\gamma)(s)$ and hence $g_n(\gamma) \to g(\gamma)$ in the finite measure equivalent to m. Thus, ϕ_γ is continuous.

3. Conversely, let $\psi : G \to Z^1(Y, \mathbb{T})$ be a continuous homomorphism. In particular, $\psi \in U(G, U(Y, \mathbb{T}))$. By the Fubini theorem for this case (see [24, Theorem 1]), there exists the isomorphism

$$F : U(G, U(Y, \mathbb{T})) \xrightarrow{\sim} U(Y, U(G, \mathbb{T}))$$

Let us clear the nature of $F(\psi)$ for $\psi \in U(G, U(Y, \mathbb{T}))$.

   (a) For a. e. $s \in Y$, the map $g \mapsto \psi(g)(s)$ (with s fixed) is a homomorphism. Indeed, the multiplication in $Z^1(Y, \mathbb{T})$ is pointwise. Hence, the map $g \mapsto \psi(g)(s)$ is a Borel homomorphism of G into $\mathbb{T}$. By [18, Theorem (22.18)], this homomorphism is continuous. Thus, $\psi(g)(s)$ is continuous with respect to g for a. e. $s \in Y$. But $(F(\psi))(s)(g) = \psi(g)(s)$, hence $(F(\psi))(s)(g)$ is continuous with respect to g. Thus, $(F(\psi))(s) \in \hat{G}$ for a. e. $s \in Y$, i. e. $F(\psi) \in U(Y, \hat{G})$.

   (b) The straight calculation shows that $F(\psi) \in Z^1(Y, \hat{G})$.

   (c) Finally, it is easy to see that $\psi = \phi_{F(\psi)}$.

4. Let $\gamma_1, \gamma_2 \in Z^1(Y, \hat{G})$ and $\gamma_1 \neq \gamma_2$. Then $F^{-1}(\gamma_1) \neq F^{-1}(\gamma_2)$, i. e. there exists $g \in G$ such that $(F^{-1}(\gamma_1))(g) \neq (F^{-1}(\gamma_2))(g)$. Hence there exists a Borel set $E \subset Y$ such that $m(E) > 0$ and $(F^{-1}(\gamma_1))(s) \neq (F^{-1}(\gamma_2))(s)$ for $s \in E$. In other words, for $s \in E$:

$$\gamma_1(s)(g) \neq \gamma_2(s)(g)$$

Thus, for such s: $g(\gamma_1)(s) = g(\gamma_1(s)) = \gamma_1(s)(g) \neq \gamma_2(s)(g) = g(\gamma_2(s)) = g(\gamma_2)(s)$.

Thus, $g(\gamma_1) \neq g(\gamma_2)$ and finally:

$$\phi_{\gamma_1}(g) = g(\gamma_1) \neq g(\gamma_2) = \phi_{\gamma_2}(g)$$

hence $\phi_{\gamma_1} \neq \phi_{\gamma_2}$.

□

NOTATION. Let $\mathrm{Aut}(\tilde{M}, I(M))$ be the set of all automorphisms of $\tilde{M}$ leaving $I(M)$ pointwise fixed.

Proposition 11.2 *The homomorphism* $\mathrm{Ad} : \mathcal{V} \to \mathrm{Aut}(\tilde{M}, I(M))$ *is a homeomorphism where* $\mathcal{V}$ *is equipped with the strong operator topology and* $\mathrm{Aut}(\tilde{M}, I(M))$ *is equipped with the topology of elementwise* σ*-weak convergence.*

PROOF. It follows from the Theorems 10.3 and 10.4 that $\mathrm{Ad} : \mathcal{V} \to \mathrm{Aut}(\tilde{M}, I(M))$ is one-to-one and onto.

1. Let $(U_\alpha)_{\alpha \in A} \subset \mathcal{V}$, $U_\alpha \to U_0$ strongly. We have to prove that for every $T \in \tilde{M}$ and every $\psi \in \tilde{M}_*$: $\psi(U_\alpha T U_\alpha^*) \to \psi(U_0 T U_0^*)$. It is enough to prove this only for $\psi \in \tilde{M}_*^+$. But $\tilde{M}$ has a cyclic and separating vector, so by [6, Theorem 2.5.31] every such ψ has a form $\psi(T) = \langle T\eta, \eta \rangle$ for some $\eta \in \tilde{H}$. Thus, it is enough to prove that for $T \in \tilde{M}$, $\eta \in \tilde{H}$:

$$\langle U_\alpha T U_\alpha^* \eta, \eta \rangle \to \langle U_0 T U_0^* \eta, \eta \rangle$$

We have: $U_\alpha \to U_0$ strongly hence $U_\alpha \to U_0$ weakly hence $U_\alpha^* \to U_0^*$ weakly hence $U_\alpha^* \to U_0^*$ strongly hence $U_\alpha^* \eta \to U_0^* \eta$ hence

$$\langle U_\alpha T U_\alpha^* \eta, \eta \rangle = \langle T U_\alpha^* \eta, U_\alpha^* \eta \rangle \to \langle T U_0^* \eta, U_0^* \eta \rangle = \langle U_0 T U_0^* \eta, \eta \rangle$$

2. Let $(U_\alpha)_{\alpha \in A} \subset \mathcal{V}$, $\mathrm{Ad}U_\alpha \to \mathrm{Ad}U_0$ elementwise σ-weakly on $\tilde{M}$. It is enough to prove that for every $\xi, \zeta \in \tilde{H}$: $\langle U_\alpha \zeta, \xi \rangle \to \langle U_0 \zeta, \xi \rangle$.

   (a) First, let $\mathrm{supp}\,\xi \subset \Gamma(\tau)$, $\tau \in \Sigma$, $\Gamma(\tau) \cap \triangle = \emptyset$. Let $\theta \in \tilde{H}$, $\mathrm{supp}\,\theta \subset \triangle$. Let $\eta = \xi + \theta$. We have: $\langle U_\alpha \pi(\tau) U_\alpha^* \eta, \eta \rangle \to \langle U_0 \pi(\tau) U_0^* \eta, \eta \rangle$.
   The straight calculation shows that

$$\langle U_\alpha \pi(\tau) U_\alpha^* \eta, \eta \rangle = \langle \zeta, U_\alpha^* \xi \rangle$$

where $\zeta(x,y) = \begin{cases} c(\tau^{-1}y, y, y)\theta(y,y) & \text{if } x = \tau^{-1}y \\ \text{arbitrary} & \text{if } x \neq \tau^{-1}y \end{cases}$.
The same is true for U_0 in place of U_α. Thus,

$$\langle \zeta, U_\alpha^* \xi \rangle = \langle U_\alpha \pi(\tau) U_\alpha^* \eta, \eta \rangle \to \langle U_0 \pi(\tau) U_0^* \eta, \eta \rangle = \langle \zeta, U_0^* \xi \rangle$$

This is true for every θ as above, so this is true for every $\zeta \in \tilde{H}$, so $U_\alpha^* \xi \to U_0^* \xi$ weakly in $\tilde{H}$.

(b) Now, let $\operatorname{supp} \xi \subset \triangle$. Then

$$\begin{aligned} (U_\alpha^* \xi)(x,y) &= U_\alpha^*(x,y)\xi(x,y) \\ &= \begin{cases} U_\alpha^*(y,y)\xi(y,y) & \text{if } x = y \\ 0 & \text{if } x \neq y \end{cases} = \begin{cases} \xi(y,y) & \text{if } x = y \\ 0 & \text{if } x \neq y \end{cases} = \xi(x,y) \end{aligned}$$

Thus, $U_\alpha^* \xi = \xi$, and the same is true for U_0 in place of U_α. Thus, $\langle \zeta, U_\alpha^* \xi \rangle = \langle \zeta, \xi \rangle \to \langle \zeta, \xi \rangle = \langle \zeta, U_0^* \xi \rangle$ in this case too.

(c) Let $\operatorname{supp} \xi \subset \bigcup_{i=1}^n \Gamma(\tau_i)$ with $\tau_i \in \Sigma$, $i = 1, \ldots, n$. By parts (2a) and (2b) of this proof, $U_\alpha \xi \to U_0 \xi$ weakly in $\tilde{H}$.

(d) Let $\xi, \zeta \in \tilde{H}$ be arbitrary, $\varepsilon > 0$. There exists $\xi_0 \in \tilde{H}$ such that $\|\xi - \xi_0\| < \varepsilon/(3\|\zeta\|)$ and $\operatorname{supp} \xi_0 \subset \bigcup_{i=1}^n \Gamma(\tau_i)$ with $\tau_i \in \Sigma$, $i = 1, \ldots, n$. By part (2d), there exists $\alpha_0 \in A$ such that for every $\alpha \geq \alpha_0$:

$$|\langle \zeta, U_\alpha^* \xi_0 \rangle - \langle \zeta, U_0^* \xi_0 \rangle| < \frac{\varepsilon}{3}$$

Then for $\alpha \geq \alpha_0$:

$$\begin{aligned} &|\langle \zeta, U_\alpha^* \xi \rangle - \langle \zeta, U_0^* \xi \rangle| \\ &\leq |\langle \zeta, U_\alpha^* \xi \rangle - \langle \zeta, U_\alpha^* \xi_0 \rangle| + |\langle \zeta, U_\alpha^* \xi_0 \rangle - \langle \zeta, U_0^* \xi_0 \rangle| \\ &\qquad\qquad + |\langle \zeta, U_0^* \xi_0 \rangle - \langle \zeta, U_0^* \xi_0 \rangle| \\ &\leq \|\zeta\| \cdot \|\xi - \xi_0\| + |\langle \zeta, U_\alpha^* \xi_0 \rangle - \langle \zeta, U_0^* \xi_0 \rangle| + \|\zeta\| \cdot \|\xi - \xi_0\| \\ &< \|\zeta\| \cdot \frac{\varepsilon}{3\|\zeta\|} + \frac{\varepsilon}{3} + \|\zeta\| \cdot \frac{\varepsilon}{3\|\zeta\|} = \varepsilon \end{aligned}$$

□

Proposition 11.3 *Let $(\Omega, \mathcal{S}, \sigma)$ be some σ-finite measure space, and let τ be a finite measure on Ω which is equivalent to σ and such that $\tau(E) \leq \sigma(E)$ for $E \in \mathcal{S}$. Let $(U_\alpha)_{\alpha \in A}$ be a bounded net of functions in $L^\infty(\Omega)$, $U_0 \in L^\infty(\Omega)$. Then (U_α) converges to U_0 strongly as operators on $L^2(\Omega)$ if and only if (U_α) converges to U_0 in measure τ (as functions on Ω).*

PROOF. In what follows, $\|\cdot\|_\sigma$ means the norm in $L^2(\Omega, \sigma)$, and $\|\cdot\|_\tau$ means the norm in $L^2(\Omega, \tau)$. Let (U_α) be bounded by M.

1. First, let $U_\alpha \to U_0$ in measure τ. We can assume that $(U_\alpha)_{\alpha \in A}$ is a sequence. By the theorem known, every subsequence of (U_α) has a subsequence converging a. e. Suppose that (U_α) do not converge to U_0 strongly. Then there

exists $\phi \in L^2(\Omega, \sigma)$ such that $(\|U_\alpha \phi - U_0 \phi\|_\sigma)$ does not converge to 0. But the last sequence is bounded, so there exists some subsequence (U_{α_n}) such that $\|U_{\alpha_n} \phi - U_0 \phi\| \to \beta \neq 0$. There exists a subsequence $(U_{\alpha_{n_m}})$ of (U_{α_n}) such that $U_{\alpha_{n_m}} \to U_0$ a. e. Then we have:

$$\|U_{\alpha_{n_m}} \phi - U_0 \phi\|_\sigma^2 = \int_\Omega |U_{\alpha_{n_m}}(\omega)\phi(\omega) - U_0(\omega)\phi(\omega)|^2 d\sigma(\omega)$$
$$= \int_\Omega |U_{\alpha_{n_m}}(\omega) - U_0(\omega)|^2 |\phi(\omega)| d\sigma(\omega) \to 0$$

because the integrand is bounded by $4M^2|\phi(\omega)|^2$ and we can use the Lebeque convergence theorem.

This is the contradiction.

2. Conversely, let $U_\alpha \to U_0$ strongly on $L^2(\sigma)$. By the property of τ, $\tau(E) \leq \sigma(E)$ for every $E \in \mathcal{M}$, so $\int_\Omega f d\tau \leq \int_\Omega f d\sigma$ for every measurable non-negative function f, so $L^2(\sigma) \subseteq L^2(\tau)$ and for $f \in L^2(\sigma)$: $\|f\|_\tau \leq \|f\|_\sigma$.

Let $\phi_0 \in L^2(\tau)$ be as follows: $\phi_0(\omega) = 1$ for $\omega \in \Omega$. Let $\phi_n \in L^2(\sigma)$ be such that $\phi_n \to \phi_0$ in $L^2(\tau)$. We have: for every $n \in \mathbb{N}$, $U_\alpha \phi_n \to U_0 \phi_n$ in $L^2(\sigma)$. By properties of τ it follows that $U_\alpha \phi_n \to U_0 \phi_n$ in $L^2(\tau)$. Furthermore, $U_0 \phi_n \to U_0 \phi_0$ in $L^2(\tau)$, and for every α: $U_\alpha \phi_n \to U_\alpha \phi_0$ in $L^2(\tau)$ uniformly with respect to $\alpha \in A$.

Let us prove that $U_\alpha \phi_0 \to U_0 \phi_0$ in $L^2(\tau)$. Let $\varepsilon > 0$. Let $N \in \mathbb{N}$ be such that for $n \geq N$, $\|\phi_n - \phi_0\|_\tau \leq (\varepsilon/3M)$. Let $\alpha_0 \in A$ be such that for $\alpha \geq \alpha_0$: $\|U_\alpha \phi_N - U_0 \phi_N\|_\tau < \varepsilon/3$. Then for $\alpha \leq \alpha_0$:

$$\|U_\alpha \phi_0 - U_0 \phi_0\|_\tau$$
$$\leq \|U_\alpha \phi_0 - U_\alpha \phi_N\|_\tau + \|U_\alpha \phi_N - U_0 \phi_N\|_\tau + \|U_0 \phi_0 - U_0 \phi_0\|_\tau$$
$$\leq \|U_\alpha\| \cdot \|\phi_N - \phi_0\|_\tau + \|U_\alpha \phi_n - U_0 \phi_n\|_\tau + \|U_0\| \cdot \|\phi_n - \phi_0\|_\tau$$
$$\leq \frac{\varepsilon}{3} + \frac{\varepsilon}{3} + \frac{\varepsilon}{3} = \varepsilon$$

Thus, $U_\alpha \phi_0 \to U_0 \phi_0$ in $L^2(\tau)$.

Let $U_\alpha \not\to U_0$ in measure τ. Then there exists $\varepsilon_0 > 0$ such that

$$\tau\{\omega \in \Omega : |U_\alpha(\omega) - U_0(\omega)| > \varepsilon_0\} \not\to 0$$

Then:

$$\|U_\alpha \phi_0 - U_0 \phi_0\|_\tau^2 = \|(U_\alpha - U_0)\phi_0\|_\tau^2$$
$$= \int_\Omega |(U_\alpha(\omega) - U_0(\omega))\phi(\omega)|^2 d\tau(\omega)$$
$$\geq \int_{\{\omega \in \Omega : |U_\alpha(\omega) - U_0(\omega)| > \varepsilon_0\}} |(U_\alpha(\omega) - U_0(\omega))\phi(\omega)|^2 d\tau(\omega)$$
$$= \int_{\{\omega \in \Omega : |U_\alpha(\omega) - U_0(\omega)| > \varepsilon_0\}} |U_\alpha(\omega) - U_0(\omega)|^2 d\tau(\omega)$$
$$\geq \varepsilon_0^2 \cdot \tau\{\omega \in \Omega : U_\alpha(\omega) - U_0(\omega) > \varepsilon_0\} \not\to 0$$

This is the contradiction.

$\square$

Theorem 11.4 *Let $\{\lambda_t\}_{t\in\mathbb{R}}$ be a one-parameter σw-continuous group of automorphisms of $\tilde{M}$ leaving $I(M)$ fixed pointwise. Then $\{\lambda_t\}$ is of the following form: $\lambda_t(T) = \exp(itE)T\exp(-itE)$, $T \in \tilde{M}$, where $E\eta\tilde{Z}$, $E = E^*$ and for a. e. $x \sim y \sim z$:*

$$E(z,y) = \alpha_{(y,x)}(E(z,x)) + c(z,x,y)E(x,y)c(z,x,y)^{-1} \tag{11.1}$$

(Here the automorphisms acting on non-bounded operators are defined as follows: $\alpha_{(y,x)}(C) = U_{(y,x)}CU_{(x,y)}$.)

In particular, $E(x,x) = 0$ for a. e. $x \in X$.

Proof. By Propositions 11.2 and 11.3, $\lambda_t = \mathrm{Ad}V_t$ with $t \in \mathbb{R}$, where $\{V_t\} \subset \mathcal{V} \cong Z^1(Y,\mathbb{T})$ is a one-parameter group, which is continuous in the measure convergence. By Theorem 11.1 with $\mathbb{R}$ in place of G, $V_t = t(\gamma)$ for some $\gamma \in Z^1(Y,\mathbb{R})$. Thus, for $s \in Y$:

$$V_t(s) = t(\gamma)(s) = t(\gamma(s)) = \exp(it\gamma(s));$$
$$V_t(x,y) = \exp(itE(x,y))$$

where $E(x,y) = \gamma(s)$, $s \in Y_{(x,y)}$. Finally, $\gamma \in Z^1(Y,\mathbb{R})$, hence

$$\gamma(z,y)(s) = \gamma(z,x)(\bar{\alpha}^z_{(y,x)}(s)) + \gamma(x,y)(\bar{c}(z,x,y)(s))$$

which implies (11.1).

□

CHAPTER 12

THE FELDMAN-MOORE-TYPE STRUCTURE THEOREM

Let $\tilde{M}$ be a von Neumann algebra on a separable Hilbert space $\tilde{H}$, M be a von Neumann algebra on a separable Hilbert space H, $I : M \to \tilde{M}$ be an injective homomorphism, $M = \int_X M(x)d\mu(x)$ be a decomposition of M into a direct integral, $H = \int_X H(x)d\mu(x)$ be the corresponding decomposition of H, $A = \int_X \mathbb{C} \cdot 1_{H(x)}d\mu(x) \cong L^\infty(X, \mu)$ be the abelian algebra of diagonal operators in M.

Definition 12.1 *With the notations above:*
$I(M)$ will be called **maximal** *in $\tilde{M}$ if*

$$\tilde{M} \cap I(A)' = I(M)$$

A subgroup N of the group of all unitary operators in $\tilde{M}$ will be called a **regularizer** *for $I(M)$ if:*

1. *N containes all unitaries from $I(M)$;*
2. *for every $U \in N$: $UI(M)U^* = I(M)$ and $UI(A)U^* = I(A)$;*
3. *there exist a map $\alpha_{(\cdot)}$ from N to $\mathrm{Aut}(M)$ such that*
   - *(a) if $U \in N$ and for some projection $P \in A$: $I(P)U \in I(M)$, then $\alpha_U|(MP) = \mathrm{id}_{(MP)}$,*
   - *(b) if $U_1, U_2 \in N$ then $\alpha_{U_1} \circ \alpha_{U_2} = \alpha_{U_1U_2}$,*
   - *(c) if $U \in N$, $a \in A$, then $I(\alpha_U(a)) = UI(a)U^*$.*

$I(M)$ will be called **regular** *in $\tilde{M}$ if it has a regularizer N which is total in $\tilde{M}$:*

$$\overline{\mathrm{lin}(N)}^w = \tilde{M}$$

$I(M)$ will be called a **Cartan subalgebra** *of $\tilde{M}$ if it is maximal, regular and there exist a faithful normal conditional expectation $E : \tilde{M} \to I(M)$.*

REMARK.

1. If $\tilde{M}$ is the crossed product of M by R as above, then $I(M)$ is a Cartan subalgebra in $\tilde{M}$. Indeed, the maximality of $I(M)$ and existence of the conditional expectation was proved. For a total regularizer of $I(M)$ we can take $\{\pi(\tau) \mid \tau \in \Sigma\}$; and for $\alpha_{\pi(\tau)}$ we can take α_τ (see the beginning of the section 3).

2. If $A = Z(M)$ then the condition $UI(A)U^* = I(A)$ follows from $UI(M)U^* = I(M)$.

The following theorem is the main result of this section.

Theorem 12.2 (The Structure Theorem) (See [15, Theorem 1].) *Let $\tilde{M}$ be a von Neumann algebra acting on a separable Hilbert space $\tilde{H}$, $M = \int_X M(x)d\mu(x)$ be another von Neumann algebra decomposed into a direct integral. Let M possess a cyclic and separating vector. Let $I : M \to \tilde{M}$ be an embedding such that $I(M)$ is a Cartan subalgebra of $\tilde{M}$.*

Then there exist: a standard equivalence relation R with countable cosets on the space (X, μ), a "2-cocycle" c on R, and a set of correspondence maps α such that $\tilde{M}$ is isomorphic to the crossed product of M by R.

PROOF. The proof of this theorem occupies the remainder of this section.

Let $I(M)$ be a Cartan subalgebra of $\tilde{M}$. Let M possess a cyclic and separating vector $\phi_0 \in H$. Then $\langle E(\cdot)\phi_0, \phi_0\rangle$ is a faithful normal state on $\tilde{M}$. So, we can assume that $\tilde{M}$ has a cyclic and separating vector $\tilde{\phi}_0$ such that $\langle T\tilde{\phi}_0, \tilde{\phi}_0\rangle_{\tilde{H}} = \langle E(T)\phi_0, \phi_0\rangle_H$ for $T \in \tilde{M}$.

Let $\tilde{J}$ be the unitary involution in the Tomita-Takesaki theory, associated with $\tilde{M}$ and $\tilde{\phi}_0$. Then $\tilde{J}\tilde{M}\tilde{J} = \tilde{M}'$.

Let $B = \tilde{J}I(A)\tilde{J}$, C be the algebra generated by $I(A)$ and B. Then C is an abelian on Neumann algebra, so $C \cong L^\infty(R, \nu)$ for some R and ν.

Proposition 12.3 (See [15, Theorem 1].) *There exists an embedding $R \hookrightarrow X \times X$ such that R becomes an equivalence relation on X with countable cosets, and ν is equivalent to a right counting measure with respect to R.*

PROOF. There are two injective homomorphisms of A into C:

$$\begin{aligned} i(a) &= I(a), & a \in A \\ j(a) &= \tilde{J}I(a)\tilde{J}, & a \in A \end{aligned}$$

These homomorphisms correspond to Borel maps π_l and π_r from R to X such that for $a \in A$, $z \in R$:

$$i(a)(z) = a(\pi_l(z)), \quad j(a)(z) = a(\pi_r(z))$$

and such that $\pi_l(\nu)$ and $\pi_r(\nu)$ are equivalent to μ.

Let $\pi : R \to X \times X$ be defined by

$$\pi(z) = (\pi_l(z), \pi_r(z)), \quad z \in R$$

π is a Borel map and it induces a *-homomorphism $\Psi : L^\infty(X \times X, \lambda) \to L^\infty(R, \nu)$ in the usual way:

$$(\Psi(f))(r) = f(\pi_l(r), \pi_r(r)), \quad f \in L^\infty(X \times X),\ r \in R$$

where λ is defined by $\lambda(E) = \nu(\pi^{-1}(E))$ for a Borel subset E of $X \times X$.

Ψ is well defined by the definition of λ. Let us prove that Ψ is onto. For every $a \in A$ let $f_a \in L^\infty(X \times X)$ be defined by $f_a(x,y) = a(x)$, where $x, y \in X$. Then for every $r \in R$:

$$(\Psi(f_a))(r) = f_a(\pi_l(r), \pi_r(r)) = a(\pi_l(r)) = i(a)(r)$$

Thus, $\Psi(f_a) = i(a)$ hence $i(a) \in \mathrm{Im}(\Psi)$. Similarly, let $g_a \in L^\infty(X \times X)$ be defined by $g_a(x,y) = a(y)$, where $x, y \in X$. Then for every $r \in R$:

$$(\Psi(g_a))(r) = g_a(\pi_l(r), \pi_r(r)) = a(\pi_r(r)) = j(a)(r)$$

Thus, $\Psi(g_a) = j(a)$ hence $j(a) \in \mathrm{Im}\Psi$. So, $I(A) \subset \mathrm{Im}\Psi$ and $B = \tilde{J}I(A)\tilde{J} \subset \mathrm{Im}\Psi$. But $I(A)$ and B generate C, so $\mathrm{Im}\Psi = C$ and Ψ is onto.

Let $K = \mathrm{Ker}\Psi$. Then K is an ideal in $L^\infty(X \times X)$; let K be defined by a projection $\chi_P \in L^\infty(X \times X)$, where $P \subset X \times X$ is Borel. Then $\Psi | L^\infty(X \times X)\chi_{P^C}$ is one-to-one. By [10, Appendice IV] it follows that π is one-to-one map from a conull subset of R onto a conull subset of P^C. Thus we can assume that $R \subset X \times X$, and for $(x,y) \in R$: $\pi_l(x,y) = x$, $\pi_r(x,y) = y$, and $\pi_r(\nu) = \mu$.

Furthermore, the map $T \mapsto \tilde{J}T^*\tilde{J}$ is a *-antiautomorphism of $B(\tilde{H})$ which carries $I(A)$ onto B and B onto $I(A)$. So it carries C onto itself and it is a *-automorphism of C. So it is induced by a Borel map $\theta : R \to R$, such that $\theta(\nu) \sim \nu$.

Let $a \in A$. We have: $j(a) = \tilde{J}i(a)^*\tilde{J}$, so

$$\begin{aligned} a(\pi_r(x,y)) &= a(y) = j(a)(x,y) = (\tilde{J}i(a)^*\tilde{J})(x,y) \\ &= (i(a))(\theta(x,y)) = a(\pi_l\theta(x,y)) \end{aligned}$$

for a. e. $(x,y) \in R$. Thus, $\pi_r = \pi_l\theta$ a. e. and because $\theta^2 = \mathrm{id}$: $\pi_l = \pi_r\theta$ a. e. Hence or a. e. $(x,y) \in R$:

$$\theta(x,y) = (\pi_l\theta(x,y), \pi_r\theta(x,y)) = (\pi_r(x,y), \pi_l(x,y)) = (y,x)$$

But θ carries R to R, so R is symmetric and $\nu \sim \nu^{-1}$ (where $d\nu^{-1}(x,y) = d\nu(y,x)$).

Before finishing the proof of Proposition 12.3 we need some consideration.

Lemma 12.4 (See [15, Proposition 3.1].) *For $T \in \tilde{M}$, $a \in A$: $\langle TI(a)\tilde{\phi}_0, \tilde{\phi}_0\rangle = \langle I(a)T\tilde{\phi}_0, \tilde{\phi}_0\rangle$.*

PROOF.

$$\begin{aligned} \langle TI(a)\tilde{\phi}_0, \tilde{\phi}_0\rangle &= \langle E(TI(a))\phi_0, \phi_0\rangle = \langle E(T)I(a)\phi_0, \phi_0\rangle \\ &= \langle I(a)E(T)\phi_0, \phi_0\rangle = \langle E(I(a)T)\phi_0, \phi_0\rangle = \langle I(a)T\tilde{\phi}_0, \tilde{\phi}_0\rangle \end{aligned}$$

□

Corollary 12.5 *$\tilde{\Delta}$ commutes with all $I(a)$ for $a \in A$.*

PROOF. This follows from the Tomita-Takesaki theory. Indeed, by [19, Proposition 9.2.14], for $a \in A$, $t \in \mathbb{R}$: $\tilde{\sigma}_t(I(a)) = I(a)$, i. e. $I(a)$ commutes with $\tilde{\Delta}^{it}$. Hence $I(a)$ commutes with $\tilde{\Delta}$.

□

Lemma 12.6 (See [15, Proposition 3.5].) *For $a \in A$: $\tilde{J}I(a)\tilde{J}\tilde{\phi}_0 = I(a^*)\tilde{\phi}_0$.*

PROOF.

$$\begin{aligned}
&\tilde{J}I(a)\tilde{J}\tilde{\phi}_0 = \tilde{J}I(a)\tilde{\phi}_0 = \tilde{J}I(a)\tilde{\Delta}^{1/2}\tilde{\phi}_0\\
&= \tilde{J}\tilde{\Delta}^{1/2}I(a)\tilde{\phi}_0 = \tilde{S}I(a)\tilde{\phi}_0 = I(a)^*\tilde{\phi}_0 = I(a^*)\tilde{\phi}_0
\end{aligned}$$

□

NOTATION. Let $\tilde{H} = \int_R H(x,y)d\nu(x,y)$ be the decomposition of $\tilde{H}$ corresponding to the Abelian algebra C.

Let as above $\triangle = \{(x,x) \mid x \in X\}$.

Lemma 12.7 (See [15, Proposition 3.6].) $\mathrm{supp}\tilde{\phi}_0 \subset \triangle$. *In particular,* $\nu(\triangle) > 0$.

PROOF. Let E be a Borel subset of X. Then:

$$i(\chi_E)\tilde{\phi}_0 = I(\chi_E)\tilde{\phi}_0 = I(\chi_E^*)\tilde{\phi}_0 = \tilde{J}I(\chi_E)\tilde{J}\tilde{\phi}_0 = j(\chi_E)\tilde{\phi}_0$$

Hence, for a. e. $(x,y) \in R$:

$$\begin{aligned}
&\chi_E(x)\tilde{\phi}_0(x,y) = i(\chi_E)(x,y)\tilde{\phi}_0(x,y) = (i(\chi_E)\tilde{\phi}_0)(x,y)\\
&= (j(\chi_E)\tilde{\phi}_0)(x,y) = j(\chi_E)(x,y)\tilde{\phi}_0(x,y) = \chi_E(y)\tilde{\phi}_0(x,y)
\end{aligned}$$

Thus, $\mathrm{supp}\,\tilde{\phi}_0 \subset (E \times E) \cup (E^C \times E^C)$. This is true for every Borel subset E of X, hence $\mathrm{supp}\,\tilde{\phi}_0 \subset \triangle$.

□

NOTATION. Let N be a total regularizer of $I(M)$ in $\tilde{M}$, $M' = \tilde{J}I(M)\tilde{J}$, $N' = \tilde{J}N\tilde{J}$.

REMARK. It follows that $U(I(M))$ and $U(M')$ are normal subgroups in N and N' respectively. (Here and below $U(B)$ is the set of all unitary operators in B.)

Proposition 12.8 *There exist subgroups $\tilde{G}$ and $\tilde{G}'$ of N and N' respectively such that:*

1. *$U(I(M)) \subset \tilde{G}$, $U(M') \subset \tilde{G}'$;*
2. *the quotient groups $\tilde{G}/U(I(M))$ and $\tilde{G}'/U(M')$ are countable;*
3. *$\tilde{G}$ and $\tilde{G}'$ are strongly dense in N and N' respectively.*

PROOF. The algebras $\tilde{M}$ and $\tilde{M}'$ have separable preduals. Hence, N and N' with the strong operator topology have countable bases. It follows from [20, Theorem I.14] that N and N' are separable. Let N_0 and N_0' be their dense countable subsets. Let $\bar{N}_0$ and $\bar{N}_0'$ be the groups generated by N_0 and N_0' respectively. Then $\bar{N}_0$ and $\bar{N}_0'$ are countable. Indeed, $\bar{N}_0$ (resp. $\bar{N}_0'$) is the set of all finite products of elements of N_0 (resp. N_0') and their inverses. Now, it is enough to take

$$\tilde{G} = \{g \cdot u \mid g \in \bar{N}_0,\ u \in U(I(M))\}, \quad \tilde{G}' = \{g' \cdot u' \mid g' \in \bar{N}_0',\ u' \in U(M')\}$$

□

NOTATION. For $\tilde{G}$ and $\tilde{G}'$ as in Proposition 12.8 above, let $G = \tilde{G}/U(I(M))$ and $G' = \tilde{G}'/U(M')$.

NOTATION. The group N acts on $I(A)$ as a group of automorphisms, and it acts on B trivially. Thus, N acts as automorphisms of C. Thus, every element $U \in N$ determines a Borel automorphism ϕ_U of X and a Borel automorphism $\tilde{\phi}_U$ of R.

Lemma 12.9 *In the notation above, for a. e. $(x,y) \in R$:*

$$\tilde{\phi}_U(x,y) = (\phi_U x, y), \quad (x,y) \in R, \quad U \in N$$

PROOF. It is enough to proof that for $f \in C \cong L^\infty(R,\nu)$: $\mathrm{Ad}(f)(x,y) = f(\phi_u x, y)$. It is enough to prove this only for $f \in I(A)$ and for $f \in B$.

First, let $f = I(a) = i(a), \quad a \in A$.

$$\begin{aligned}\mathrm{Ad}U(f)(x,y) &= \mathrm{Ad}U(i(a))(x,y) = i(\overline{\mathrm{Ad}}U(a))(x,y)\\ &= (\overline{\mathrm{Ad}}U(a))(x) = a(\phi_U x) = i(a)(\phi_U x, y) = f(\phi_U x, y)\end{aligned}$$

where $\overline{\mathrm{Ad}}U \in \mathrm{Aut}(A)$ is defined by: $i(\overline{\mathrm{Ad}}U(a)) = \mathrm{Ad}U(i(a))$ for $a \in A$.

Now let $f = j(a) = \tilde{J}I(a)\tilde{J}$, $a \in A$.

$$\begin{aligned}(\mathrm{Ad}Uf)(x,y) &= \mathrm{Ad}U(j(a))(x,y)\\ &= j(a)(x,y) = a(y) = j(a)(\phi_u x, y) = f(\phi_U x, y)\end{aligned}$$

□

NOTATION. Similarly, every $U' \in N'$ determines a Borel automorphism $\psi_{U'}$ of X and a Borel automorphism $\tilde{\psi}_{U'}$ of R such that $\tilde{\psi}_{U'}(x,y) = (x, \psi_{U'} y)$ for $(x,y) \in R$.

Lemma 12.10 *In the notation above, for a. e. $(x,y) \in R$:*

$$\tilde{\phi}_{U'}(x,y) = (\phi_{U'} x, y), \quad (x,y) \in R, \quad U \in N$$

PROOF. The proof is just the repetition of the proof of Lemma 12.9.

□

Lemma 12.11 *For* $U \in U(I(M))$: $\phi_U = \mathrm{id}_X$. *For* $U' \in U(M')$: $\phi_{U'} = \mathrm{id}_X$.

PROOF. If $U \in U(I(M))$ then U commutes with elements from A, so it defines the trivial map on X. The same is true for $U' \in U(M')$.

□

The maps $U \mapsto \phi_U$ and $U' \mapsto \psi_{U'}$ defined above are group homomorphisms from N and N' respectively to Borel automorphisms of X. By Lemma 12.11 these homomorphisms depend only on cosets modulo $U(I(A))$ and $U(B)$ respectively, i. e. they are indeed the homomorphisms of G and G' respectively. Thus, G and G' act as transformation groups on X preserving the class of μ, and $G \times G'$ acts on R preserving the class of ν.

NOTATION. We will denote the action of the element $g \in G$ by ϕ_g, the action of the element $g' \in G'$ — by $\psi_{g'}$.

Lemma 12.12 *For* $U \in N$: $\mathrm{supp}\,(U\tilde{\phi}_0) \subseteq \Gamma(\phi_U)$. *For* $U' \in N'$: $\mathrm{supp}\,(U'\tilde{\phi}_0) \subseteq \Gamma(\psi_{U'^{-1}})$.

PROOF. Let $P = \chi_{\Gamma(\phi_U)} \in L^\infty(R) \cong C$. Then

$$(U^{-1}PU)(x,y) = P(\phi_{U^{-1}}x, y)$$
$$= \begin{cases} 0 & \text{if } y \neq \phi_U\phi_{U^{-1}}x \\ 1 & \text{if } y = \phi_U\phi_{U^{-1}}x \end{cases} = \begin{cases} 0 & \text{if } y \neq x \\ 1 & \text{if } y = x \end{cases} = \chi_\triangle(x,y)$$

Hence $U^{-1}PU\tilde{\phi}_0 = \tilde{\phi}_0$ and $PU\tilde{\phi}_0 = U\tilde{\phi}_0$. Thus, $\mathrm{supp}\,(U\tilde{\phi}_0) \subseteq \Gamma(\phi_U)$.
Similarly, let $P' = \chi_{\Gamma(\psi_{U'^{-1}})}$. Then

$$(U'^{-1}P'U')(x,y) = P'(x, \psi_{U^{-1}}y)$$
$$= \begin{cases} 0 & \text{if } \psi_{U'^{-1}}y \neq \psi_{U'^{-1}}x \\ 1 & \text{if } \psi_{U'^{-1}}y = \psi_{U'^{-1}}x \end{cases} = \begin{cases} 0 & \text{if } y \neq x \\ 1 & \text{if } y = x \end{cases} = \chi_\triangle(x,y)$$

Hence $U'^{-1}P'U'\tilde{\phi}_0 = \tilde{\phi}_0$ and $P'U'\tilde{\phi}_0 = U'\tilde{\phi}_0$.

□

Corollary 12.13 $\nu(R \setminus \bigcup_{g\in G}\Gamma(\phi_g)) = 0$.

PROOF. Let $E = R \setminus \bigcup_{g\in G}\Gamma(\phi_g)$. Let $\nu(E) > 0$, $P = \chi_E$, $\xi \in \mathrm{Im}P$. Then by Lemma 12.12, for every $U \in N$:

$$\langle \xi, U\tilde{\phi}_0 \rangle = 0$$

(because $\xi \in P\tilde{H}$, $U\tilde{\phi}_0 \in (1-P)\tilde{H}$). But $\tilde{M}$ is generated by N, so $\langle \xi, T\tilde{\phi}_0 \rangle = 0$ for every $T \in \tilde{M}$. This contradicts the fact that $\tilde{\phi}_0$ is cyclic for $\tilde{M}$.

□

Thus we can assume that

$$R = \bigcup_{g \in G} \Gamma(\phi_g) \tag{12.1}$$

Corollary 12.14 *R is an equivalence relation with countable cosets.*

PROOF. It remains only to prove the transitivity. But if $(x, y), (y, z) \in R$, then $y = \phi_{g_1} x$, $z = \phi_{g_2} y$, so $z = \phi_{g_2 g_1} x$ and hence $(x, z) \in R$.

□

Lemma 12.15 $\tilde{M} \cap C = I(A)$, $\quad \tilde{M}' \cap C = B$.

PROOF. It is clear that $I(A) \subset \tilde{M} \cap C$. Now, let p be a projection in $\tilde{M} \cap C$. We have: $C \cong L^\infty(R, \nu)$, so p is the operator of multiplication by indicator of a Borel subset E of R. Let $F = \{x \in X \mid (x, x) \in E\}$. We have: $\operatorname{supp} \tilde{\phi}_0 \subset \triangle$, so $p\tilde{\phi}_0 = I(\bar{p})$ where $\bar{p} \in A$ is the operator of multiplication by indicator function of F. But $\tilde{\phi}_0$ is separating for $\tilde{M}$, so $p = I(\bar{p}) \in I(A)$.

To prove the second equation it is enough to adjoint two parts of the first equation by $\tilde{J}$.

□

Lemma 12.16 (See [15, Proposition 3.2].) *The subalgebra of C of G invariants is B; The subalgebra of C of G' invariants is $I(A)$.*

PROOF. Let $f \in C$ be fixed by G. Then $UfU^* = f$ for every $U \in \tilde{G}$. By continuity, this is true for every $U \in N$. But $\tilde{M}$ is generated by N, so f commutes with $\tilde{M}$, i. e. $f \in \tilde{M}'$. Thus, $f \in \tilde{M}' \cap C = B$ by Lemma 12.15.

The second part is analogous.

□

The measure ν on $R \subset X \times X$ can be disintegrated by the projection π_l or π_r to the first or second coordinate. The disintegration of ν with respect to π_r gives the base measure μ and fiber measures β_y on X. Similarly, disintegration with respect to π_l gives the measure μ and fiber measures γ_x on X such that

$$\nu = \int_X \beta_y d\mu(y) = \int_X \gamma_x d\mu(x)$$

Lemma 12.17 (See [15, Proposition 3.4].) *Almost all of the fiber measures β_y, $y \in X$ (respectively γ_x, $x \in X$) are quasi-invariant under G (respectively G').*

PROOF. ν is quasi-invariant under the action of G on $R \subset X \times X$, and the projection π_r commutes with this action. Hence by [4, Ch. II, Proposition 2.6] β_y are quasi-invariant under G.

The part concerning γ_x and G' is analogous.

□

Lemma 12.18 (See [15, Proposition 3.7].) *The fiber measures β_y satisfy the property $\beta_y(\{y\}) > 0$ for a. e. $y \in X$. Analogically, $\gamma_x(\{x\}) > 0$ for a. e. $x \in X$.*

PROOF. Let $\beta_y(\{y\}) = 0$ for $y \in E$, $\mu(E) > 0$. Let $P = \chi_E \in A$. Then $I(P) \in \tilde{M}$; $I(P)(x, y) = P(x) = \chi_E(x)$, so $(I(P)\tilde{\phi}_0)(x, y) = \chi_E(x)\tilde{\phi}_0(x, y)$.

$$
\begin{aligned}
&\|I(P)\tilde{\phi}_0\|^2 = \int_R \|(I(P)\tilde{\phi}_0)(x, y)\|^2 d\nu(x, y) \\
&= \int_R \|\chi_E(x)\tilde{\phi}_0(x, y)\|^2 d\nu(x, y) \\
&= \int_X \left(\int_X \|\chi_E(x)\tilde{\phi}_0(x, y)\|^2 d\beta_y(x)\right) d\mu(y) \\
&= \int_X \left(\|\chi_E(x)\tilde{\phi}_0(y, y)\|^2 \beta_y(\{y\})\right) d\mu(y) \\
&= \int_X \chi_E(x)\|\tilde{\phi}_0(y, y)\|^2 \beta_y(\{y\}) d\mu(y) \\
&= \int_E \|\tilde{\phi}_0(y, y)\|^2 \beta_y(\{y\}) d\mu(y) = \int_E 0 d\mu(y) = 0
\end{aligned}
$$

But $\tilde{\phi}_0$ is separating for $\tilde{M}$ — this is a contradiction.

□

Lemma 12.19 (See [15, Proposition 3.8].) *For almost all $y \in X$, β_y is a purely atomic measure which gives positive mass to exactly the points of the G-orbit of y, and similarly for γ_x.*

PROOF. By Corollary 12.14, $R = \bigcup_{g \in G} \Gamma(\phi_g)$ up to a null set. So, for $x \in X$ the coset $R(x) = \{y \in X \mid (x, y) \in R\}$ equals to $\{y \in X \mid$ there exists $g \in G$ such that $y = \phi_g x\}$ i. e. $R(x)$ equals the G-orbit of x.

The measure ν is concentrated on R. So for a. e. $y \in X$ the measure β_y is concentrated on $R(y)$ — the G-orbit of y. Finally, let z belong to the G-orbit of y, i. e. $z = g(y)$ for some $g \in G$. Then from the G-inariance of β_y (Lemma 12.17) and from the fact that $\beta_y(\{y\}) > 0$ (Lemma 12.18) it follows that $\beta_y(\{z\}) > 0$.

□

Thus, ν is equivalent to the right counting measure (see [14, Theorem 2]). We will replace ν by the right counting measure continuing to denote it by ν.

This completes the proof of Proposition 12.3.

□

NOTATION. Let Σ be the set of all Borel automorphisms of X with graphs in R.

Lemma 12.20 *For every $\rho \in \Sigma$ there exists (not unique!) unitary operator $U_\rho \in \tilde{M}$ such that:*

1. $U_\rho I(M) U_\rho^* = I(M), \quad U_\rho I(A) U_\rho^* = I(A)$
2. *The Borel automorphism of X defined by* $\mathrm{Ad}U \in \mathrm{Aut}(I(A))$ *coincides with ρ.*

PROOF. Let $\rho \in \Sigma$. $\Gamma(\rho) \subset R = \bigcup_{g\in G} \Gamma(\phi_g)$, and G is countable. Let $G = \{g_i\}_{i=1}^{\infty}$. For every $g_i \in G$ there exists (by the definition of G) some $U_{g_i} \in N$ such that $\phi_{g_i} = \phi_{U_{g_i}}$. Now let us define:

$$P_i = \{x \in X \mid \rho x = \phi_{g_i} x\}, \quad i = 1, 2, \ldots$$

Then $\bigcup_{i=1}^{\infty} P_i = X$. Let us define:

$$Q_1 = P_1, \quad Q_i = P_i \setminus \bigcup_{k=1}^{i-1} P_k, \quad i \geq 2$$
$$S_i = \chi_{Q_i} \in A, \quad i = 1, \ldots$$

Then $S_i S_j = 0$ for $i \neq j$. Let $U = \sum_{i=1}^{\infty} I(S_i) U_{g_i}$. Then U is unitary, $U \in \tilde{M}$. For $a \in M$:

$$U^* I(a) U = \sum_{i=1}^{\infty} U_{g_i}^* I(S_i) I(a) \sum_{j=1}^{\infty} I(S_j) U_{g_j}$$
$$= \sum_{i=1}^{\infty} U_{g_i}^* I(aS_i) U_{g_i} \in I(M)$$

For $a \in A$:

$$U^* I(a) U = \sum_{i=1}^{\infty} U_{g_i}^* I(aS_i) U_{g_i}$$
$$= \sum_{i=1}^{\infty} I((aS_i) \circ \phi_{g_i^{-1}}) = \sum_{i=1}^{\infty} I((aS_i) \circ \rho^{-1}) = I(a \circ \rho^{-1})$$

by the choise of S_i, $i = 1, \ldots$. Thus, U^* defines the automorphism ρ^{-1}, hence U defines the automorphism ρ.

□

Corollary 12.21 *For $f \in C$, $\rho \in \Sigma$, a. e. $(x, y) \in R$:*

$$(U_\rho f U_\rho^*)(x, y) = f(\rho x, y)$$

PROOF. The proof is just repetition of Proof of Lemma 12.9.

□

Corollary 12.22 *Let $U'_\rho = \tilde{J} U_\rho \tilde{J}$. Then for $f \in C$, a. e. $(x, y) \in R$:*

$$(U'_\rho f U'^*_\rho)(x, y) = f(x, \rho y)$$

PROOF. The proof is the same as for U_ρ: checking for $f \in I(A)$ and for $f \in B$.

□

NOTATION. For all operators which are diagonalisable with respect to the decomposition $\tilde{H} = \int_X H(x, y) d\mu(x, y)$, we will denote their coordinates by square parentheses. Thus, we will write:

$$T = \int_R T[x, y] d\nu(x, y)$$

12.1 Structure of operators in N

Let $U \in N$.

The measure ν on R is right countable, hence $d\nu(\tilde{\phi}_U(x,y)) = d\nu(\phi_U x, y) = d\nu(x,y)$. By the definition of $\tilde{\phi}_U$ after Proposition 12.8, for $c \in C$, a. e. $(x,y) \in R$:

$$(UcU^*)(x,y) = c(\tilde{\phi}_U(x,y))$$

By [16, Proposition 4.1] this implies that there exist unitary operators $U_{(x,y)}$ acting from $H(x,y)$ onto $H(\tilde{\phi}_U^{-1}(x,y)) = H(x,y)$ such that for $f \in \tilde{H}$, a. e. $(x,y) \in R$:

$$(Uf)(x,y) = U_{(\phi_U x, y)} f(\phi_U x, y)$$

NOTATION. Let $U[x,y] = U_{(\phi_U x, y)}$, $(x,y) \in R$. Thus, for $U \in N$, a. e. $(x,y) \in R$:

$$(Uf)(x,y) = U[x,y] f(\phi_U x, y)$$

REMARK. By the definition of the map α, for every $U \in I(M)$: $\alpha_U = \text{id}$. Thus, for $U \in \tilde{G}$, α_U depends only on the coset of U in $G = \tilde{G}/U(I(M))$. We will denote it by α_g, $g \in G$.

By the definitions, for $U \in g$: $\alpha_U = \alpha_g$ and $\phi_U = \phi_g$.

Lemma 12.23 *For $F \subset X$: $\overline{\text{Ad}}U(\chi_F) = \chi_{\phi_U^{-1}F}$.*

PROOF.

$$(\overline{\text{Ad}}U(\chi_F))(x) = \chi_F(\phi_U x)$$
$$= \begin{cases} 1 & \text{if } \phi_U x \in F \\ 0 & \text{if } \phi_U x \notin F \end{cases} = \begin{cases} 1 & \text{if } x \in \phi_U^{-1} F \\ 0 & \text{if } x \notin \phi_U^{-1} F \end{cases} = \chi_{\phi_U^{-1}F}(x)$$

□

12.2 The maps $\alpha^x_{(y,z)}$ and $\beta^x_{(y,z)}$

Lemma 12.24 *The element α_U depends only on ϕ_U.*

PROOF. Let $U_1, U_2 \in N$ and $\phi_{U_1} = \phi_{U_2}$. Then $\phi_{U_1^{-1}U_2} = \text{id}$. Thus, $U_1^{-1}U_2$ commutes with $I(A)$. Hence, $U_1^{-1}U_2 \in \tilde{M} \cap I(A)' = I(M)$. By the property (3a) of $\alpha_{(\cdot)}$ (see Definition 12.1), $\alpha_{U_1^{-1}U_2} = \text{id}$ and so $\alpha_{U_1} = \alpha_{U_2}$.

□

NOTATION. We will write α_{ϕ_U} in place of α_U.

Elements of the algebra $I(M)$ commute with elements of C, so every $I(a)$, $a \in M$, is diagonalisable with respect to the decomposition $\tilde{H} = \int_R H(x,y)d\nu(x,y)$.

NOTATION. Let $I(a) = \int_R I(a)[x,y]d\nu(x,y)$ be the decomposition of $I(a)$, $a \in M$. (See Notation in the beginning of Subsection 12.1.)

Let for every $(x,y) \in R$:

$$M(x,y) = \{I(a)[x,y] \mid a \in M\}$$

Let for every $x \sim y \sim z$ the maps $\alpha^x_{(y,z)} : M(x,z) \to M(x,y)$ and $\beta^x_{(y,z)} : M(z,x) \to M(y,x)$ be defined as follows:

$$\begin{array}{rcl} \alpha^x_{(y,z)}(I(a)[x,z]) & = & I(a)[x,y] \\ \beta^x_{(y,z)}(I(a)[z,x]) & = & I(\alpha_g(a))[y,x] \end{array}$$

where $g \in G$, $y = \phi_g^{-1}z$.

The maps $\alpha^x_{(y,z)}$ and $\beta^x_{(y,z)}$ are well defined a. e. on every set of positive measure.

Theorem 12.25 *The diagram*

$$\begin{array}{ccc} M(x,z) & \xrightarrow{\alpha^x_{(y,z)}} & M(x,y) \\ \Big\downarrow \beta^z_{(v,x)} & & \Big\downarrow \beta^y_{(v,x)} \\ M(v,z) & \xrightarrow{\alpha^v_{(y,z)}} & M(v,y) \end{array}$$

is commutative for a. e. $x \in X$, $y \sim z \sim v \sim x$.

PROOF. Let $\hat{a} \in M(x,z)$, $\hat{a} = I(a)[x,z]$. Then

$$\alpha^x_{(y,z)}(\hat{a}) = I(a)[x,y];$$
$$\beta^y_{(v,x)}(\alpha^x_{(y,z)}(\hat{a})) = \beta^y_{(v,x)}(I(a)[x,y]) = I(\alpha_g(a))[v,y]$$

where $v = \phi_g^{-1}x$.

On the other hand,

$$\beta^z_{(v,x)}(\hat{a}) = I(\alpha_g(a))[v,z];$$
$$\alpha^v_{(y,z)}(\beta^z_{(v,x)}(\hat{a})) = \alpha^v_{(y,z)}(I(\alpha_g(a))[v,z]) = I(\alpha_g(a))[v,y]$$

□

NOTATION. Let $\gamma : M \to \int_X M(x,x)d\mu(x)$ be the natural correspondence:

$$\gamma(a) = I(a)P_\triangle$$

where $a \in M$, $P_\triangle$ is the projection to the diagonal $\triangle$. This γ is an automorphism.

Let $\gamma_x(a(x)) = I(a)(x,x)$, $a \in M$. Then γ_x is a. e. well defined, $\gamma_x : M(x) \to M(x,x)$ and $\gamma = \int_X \gamma_x d\mu(x)$.

Let $\beta(x,y) = (\gamma_y)^{-1} \circ \beta^y_{(y,x)}$, $\beta = \int_R \beta(x,y)d\nu(x,y)$. (Note: $M(x,y) \overset{\beta^y_{(y,x)}}{\to} M(y,y) \overset{\gamma_y^{-1}}{\to} M(y)$.)

The map β is the isomorphism transforming the von Neumann algebra $\int_R M(x,y)d\nu(x,y)$ onto the algebra $\int_R M(y)d\nu(x,y)$. Both of these algebras possess cyclic and separating vectors, so β is spatial. Hence, β is implemented by a unitary operator U_β so that $\beta = \mathrm{Ad}U_\beta$. Thus, the isomorphism β may be extended to the isomorphism $\mathrm{Ad}U_\beta$ from the algebra $B(\tilde{H})$ onto $B(\tilde{\tilde{H}})$ where $\tilde{\tilde{H}} = \int_R H(y)d\nu(x,y)$.

Lemma 12.26 *1.* $\alpha^x_{(y,z)} \circ \alpha^x_{(z,v)} = \alpha^x_{(y,v)}$;

2. $\beta^x_{(y,z)} \circ \beta^x_{(z,v)} = \beta^x_{(y,v)}$.

PROOF.

1. Let $\hat{a} \in M(x,v)$, $\hat{a} = I(a)[x,v]$, $a \in M$.

$$\begin{aligned} &\alpha^x_{(y,z)}(\alpha^x_{(z,v)}(\hat{a})) = \alpha^x_{(y,z)}(I(a)[x,z]) \\ &= I(a)[x,y] = \alpha^x_{(y,v)}(I(a)[x,v]) = \alpha^x_{(y,v)}(\hat{a}) \end{aligned}$$

2. Let $\hat{b} \in M(v,x)$, $\hat{b} = I(b)[v,x]$, $b \in M$, $z = \phi_g x$, $y = \phi_h z$, $g,h \in G$.

$$\begin{aligned} &\beta^x_{(y,z)}(\beta^x_{(z,v)}(\hat{b})) = \beta^x_{(y,z)}(I(\alpha_g(b))[z,x]) \\ &= I(\alpha_h(\alpha_g(b)))[y,x] = I(\alpha_{hg}(b))[y,x] = \beta^x_{(y,v)}(\hat{b}) \end{aligned}$$

since $y = \phi_{hg} z$.

□

NOTATION. Let us define for $(y,z) \in R$:

$$\alpha_{(y,z)} = (\gamma_y)^{-1} \circ \beta^y_{(y,z)} \circ \alpha^z_{(y,z)} \circ \gamma_z$$

Then $\alpha_{(y,z)}$ is an isomorphism from $M(z)$ onto $M(y)$, and $\alpha_{(z,z)} = \mathrm{id}_{M(z)}$. Indeed:

$$M(z) \overset{\gamma_z}{\to} M(z,z) \overset{\alpha^z_{(y,z)}}{\to} M(z,y) \overset{\beta^y_{(y,z)}}{\to} M(y,y) \overset{\gamma_y^{-1}}{\to} M(y)$$

Lemma 12.27 *The diagram*

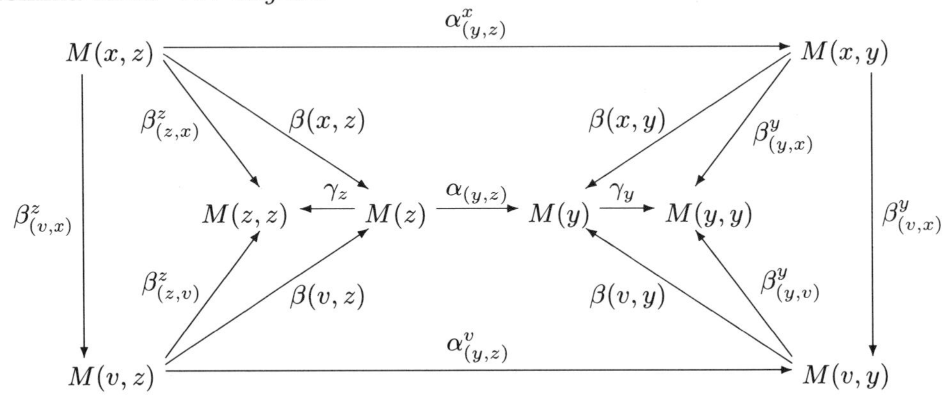

is a commutative diagram of isomorphisms.

PROOF. The right triangle and the left triangle are commutative by Lemma 12.26, part 2. Four middle triangles are commutative by the definition of $\beta(x,z)$, $\beta(x,y)$ etc. The external rectangle is commutative by Theorem 12.25. It remains only to prove the commutativity of trapezes. It is enough to check only the top trapeze. For this, let $v = z$. Then the bottom trapeze is commutative by the definition of $\alpha_{(y,z)}$. So the top trapeze is commutative too.

□

Theorem 12.28 *For* $a \in M$:

$$\beta(I(a)) = \int_R \alpha_{(y,x)}(a(x))d\nu(x,y)$$

PROOF. We have to prove that

$$\beta(x,y)(I(a)[x,y]) = \alpha_{(y,x)}(a(x))$$

Take $z = x$ in Lemma 12.27. Then:

$$\beta(x,y)^{-1} \circ \alpha_{(y,x)} = \alpha^x_{(y,x)} \circ \beta^x_{(x,x)} \circ (\gamma_x)^{-1} = \alpha^x_{(y,x)} \circ (\gamma_x)^{-1};$$

$$\beta(x,y)^{-1} \circ \alpha_{(y,x)}(a(x)) = \alpha^x_{(y,x)}(\gamma_x^{-1}(a(x)))$$

$$= \alpha^x_{(y,x)}(I(a)[x,x]) = I(a)[x,y]$$

□

This theorem means that the operator $\beta(I(a))$ acts in the space $\tilde{\tilde{H}}$ as follows:

$$(\beta(I(a))f)(x,y) = \alpha_{(y,x)}(a(x))f(x,y), \quad f \in \tilde{\tilde{H}}$$

Every isomorphism $\alpha_{(y,x)}$ is spatial hence it may be extended to the isomorphism from $B(H(x))$ onto $B(H(y))$.

Theorem 12.29 *For $y \sim z \sim x$: $\alpha_{(y,z)} \circ \alpha_{(z,x)} = \alpha_{(y,x)}$.*

PROOF. Consider the diagram:

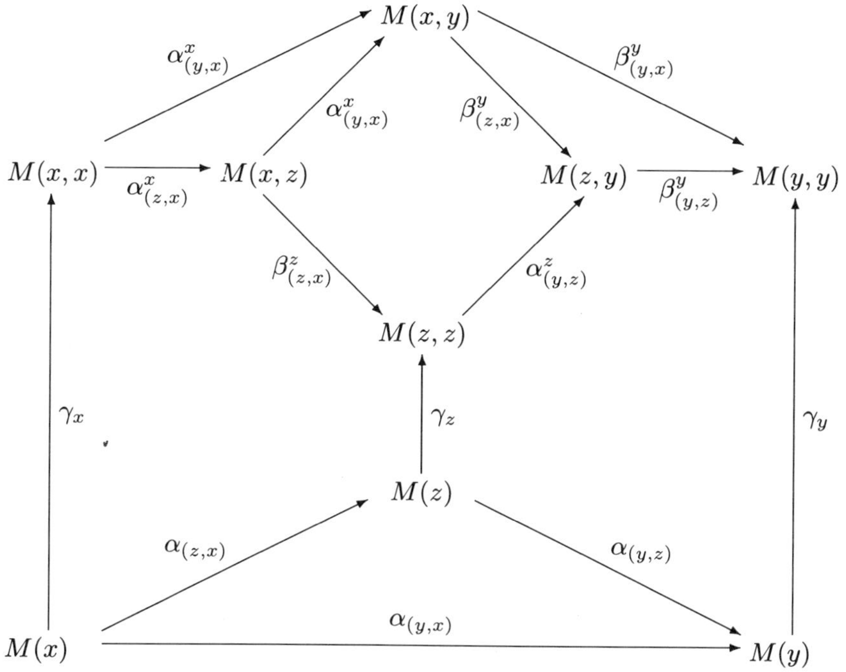

The top triangles (left and right) are commutative by Lemma 12.26. The central rhombus is commutative by Theorem 12.25. The middle pentagons (left and right) and the external pentagon are commutative by the definitions of $\alpha_{(z,x)}$, $\alpha_{(y,z)}$ and $\alpha_{(y,x)}$ respectively. Hence, the bottom triangle is commutative too.

□

12.3 Coordinate representation of elements of $\tilde{M}_0$

Let $U \in N$. By Notation at the beginning of Subsection 12.1:

$$(Uf)(x,y) = U[x,y]f(\phi_U x, y)$$

where $\phi_U : X \to X$ is a Borel automorphism, $U[x,y]$ is a unitary operator acting from $H(\phi_U x, y)$ onto $H(x,y)$, $f \in \tilde{H}$.

NOTATION. Let $\beta(U)[x,y] = \beta(x,y)(U[x,y])$, where $(x,y) \in R$.

Let also $\beta(T)[x,y] = \beta(x,y)(T[x,y])$, where T is a diagonalisable operator on $\tilde{H}$, $(x,y) \in R$.

Then the operator $\beta(U)$ has a form

$$(\beta(U)f)(x,y) = \beta(U)[x,y]f(\phi_U x, y)$$

where $f \in \tilde{\tilde{H}}$, $\beta(U)[x,y]$ is a unitary operator on the space $H(y)$ (since both of the spaces $H(\phi_U x, y)$ and $H(x,y)$ go to $H(y)$).

Lemma 12.30 *If $U \in N$, $\phi_U x = x$ for $x \in E$, where E is a Borel subset of X, then $I(\chi_E)U = UI(\chi_E) \in I(M)$, and $\alpha_{(x,y)}(\beta(U)[x,y])$ does not depend on $y \in X$ for $x \in E$, $y \sim x$.*

PROOF. $UI(\chi_E)U^* = I(\chi_{\phi_U^{-1}E}) = I(\chi_E)$, so $UI(\chi_E) = I(\chi_E)U$. Now, for every Borel $F \subseteq X$:

$$\begin{aligned} I(\chi_F)I(\chi_E)U &= I(\chi_{F\cap E})U = UU^*I(\chi_{F\cap E})U \\ &= UI(\chi_{\phi_U(F\cap E)}) = UI(\chi_{F\cap E}) \\ &= UI(\chi_E)I(\chi_F) = I(\chi_E)UI(\chi_F) \end{aligned}$$

Thus, $I(\chi_E)U$ commutes with every $I(\chi_F)$, where F is a Borel subset of X. Hence $I(\chi_E)U$ commutes with $I(A)$, and $I(\chi_E)U \in I(M)$ (because $I(M) = \tilde{M} \cap I(A)'$). Write: $I(\chi_E)U = I(a)$, $a \in M$. Then by Theorem 12.28:

$$\beta(I(\chi_E)U)[x,y] = \alpha_{(y,x)}(a(x))$$

or:

$$\alpha_{(x,y)}(\beta(I(\chi_E)U)[x,y]) = a(x)$$

But $\beta(I(\chi_E))[x,y] = \chi_E(x) \cdot 1_{H(y)}$. So, for a. e. $x \in E$, $f \in \tilde{\tilde{H}}$:

$$\begin{aligned} &(\beta(I(\chi_E)U)f)(x,y) = \beta(I(\chi_E))[x,y](\beta(U)f)(x,y) \\ &= (\beta(U)f)(x,y) = \beta(U)[x,y]f(\phi_U x, y) = \beta(U)[x,y]f(x,y) \end{aligned}$$

Thus, $\beta(I(\chi_E)U)[x,y] = \beta(U)[x,y]$, so $\alpha_{(x,y)}(\beta(I(\chi_E)U)[x,y]) = a(x)$ — does not depend on y.

□

Lemma 12.31 *There exists such a Borel countable partition T of R that:*

1. *for every $\tau \in T$ there exists $U_\tau \in N$ such that $\tau \subseteq \Gamma(\phi_{U_\tau})$;*
2. *$\triangle \in T$;*
3. *for every $\tau \in T$: $\tau^{-1} \in T$ too. Here $\tau^{-1} = \{(y,x) \mid (x,y) \in \tau\}$.*

PROOF. By the assumption (12.1) after Corollary 12.13, $R = \bigcup_{g\in G} \Gamma(\phi_g)$. Let $G = \{g_i\}_{i=1}^{\infty}$ and $g_1 = e$ (the group unit in G). Let $\sigma_1 = \Gamma(g_1) = \triangle$, and for $i \geq 2$: $\sigma_i = \Gamma(g_i) \setminus \bigcup_{j=1}^{i-1} \Gamma(g_j)$. Then σ_i are disjoint and $\bigcup \sigma_i = R$. Finally, let

$$T = \{\sigma_i \cap \sigma_j \mid i,j = 1,2,\ldots\}$$

All the properties requested are satisfied.

□

We choose U_τ for $\tau \in T$ such that $U_{\tau^{-1}} = U_\tau^{-1}$ for all $\tau \in T$.

NOTATION. Let us define $c(x,y,z) \in B(H(z))$ like in the commutative case in [15]:

$$c(x,y,z) = \beta(U_\tau)[x,z]$$

where $\tau \in T$ is such that $y = \tau x$.

Then $c(x,y,z)$ is unitary, and by the properties of T and U_τ:

$$c(x,y,z) = c(y,x,z)^{-1} \quad \text{and} \quad c(x,x.z) = 1_{H(z)} \tag{12.2}$$

NOTATION. Let $\tilde{M}_0$ be the algebra (not closed!) generated by $\beta(I(M))$ and $\beta(N)$.

The algebra $\tilde{M}$ is generated by N as a von Neumann algebra, so $\beta(\tilde{M})$ is the (weak) closure of $\tilde{M}_0$.

Theorem 12.32 (See [15, Proposition 3.16].) *For every $V \in \tilde{M}_0$, a. e. $(x,y) \in R$ there exists an operator $V(x,y) \in M(y)$, such that for every $f \in \tilde{\tilde{H}}$:*

$$(Vf)(x,z) = \sum_{y\sim x} \alpha_{(z,y)}(V(x,y))c(x,y,z)f(y,z) \tag{12.3}$$

PROOF.

1. Suppose first that $V \in \beta(N)$. For every $\tau \in T$ let $E_\tau = \{x \in X \mid \phi_V x = \tau x\}$. (Here $\phi_V = \phi_{\beta^{-1}(V)}$.) Then $\dot{\bigcup}_{\tau\in T} E_\tau = X$ is a disjoint union. Let us first prove the theorem for every $V_\tau = \beta(I(\chi_{E_\tau}))V$, where $\tau \in T$. Of course this is not trivial only if $\mu(E_\tau) > 0$. By Lemma 12.30, $\beta(I(\chi_{E_\tau}))V\beta(U_\tau^{-1}) \in \beta(I(M))$. Let $\beta(I(\chi_{E_\tau}))V\beta(U_\tau^{-1}) = \beta(I(a_\tau))$, where $a_\tau \in M$. Then $\beta(I(\chi_{E_\tau}))V = \beta(I(a_\tau))\beta(U_\tau)$. Let $f \in \tilde{\tilde{H}}$.

$$\begin{aligned}
&(V_\tau f)(x,z) = (\beta(I(\chi_{E_\tau}))Vf)(x,z) = (\beta(I(a_\tau))\beta(U_\tau)f)(x,z)\\
&= \beta(I(a_\tau))[x,z](\beta(U_\tau))f)(x,z) = \alpha_{(z,x)}(a_\tau(x))\beta(U_\tau)[x,z]f(\tau x,z)\\
&\text{(by Theorem 12.28)}\\
&= \alpha_{(z,x)}(a_\tau(x))c(x,\tau x,z)f(\tau x,z)\\
&= \alpha_{(z,\tau x)}(\alpha_{(\tau x,x)}(a_\tau(x)))c(x,\tau x,z)f(\tau x,z)\\
&= \textstyle\sum_{y\sim x} \alpha_{(z,y)}(V_\tau(x,y))c(x,y,z)f(y,z)
\end{aligned}$$

where $V_\tau(x,y) = \begin{cases} \alpha_{(\tau x,x)}(a_\tau(x)) & \text{if } y = \tau x \\ 0 & \text{otherwise} \end{cases}$. Thus, the theorem is proved for V_τ.

Now, $\sum_{\tau\in T} \chi_{E_\tau} = 1_H$, hence $V = \sum_{\tau\in T} V_\tau$ where the sum is a strong limit of the sequence of partial sums. In other words: if $T = \{\tau_i\}_{i=1}^\infty$ then $V =$ s-$\lim_{n\to\infty} \sum_{i=1}^n V_{\tau_i}$, i. e. for $f \in \tilde{\tilde{H}}$: $Vf = \lim_{n\to\infty} \sum_{i=1}^n V_{\tau_i} f$. By [10, Ch. II,

§1, n. 5, Proposition 5(ii)] there exists a subsequence $(n_k)_{k=1}^{\infty}$ of $\mathbb{N}$ such that $(Vf)(x,z) = \lim_{k\to\infty} \sum_{i=1}^{n_k} (V_{\tau_i} f)(x,z)$ a. e. on R. We have:

$$(V_{\tau_i} f)(x,z) = \alpha_{(z,\tau_i x)}(\alpha_{(\tau_i x,x)}(a_{\tau_i}(x)))c(x,\tau_i x,z)f(\tau_i x,z)$$

So,

$$\begin{aligned}
&\sum_{i=1}^{n_k}(V_{\tau_i} f)(x,z)\\
&= \sum_{i=1}^{n_k} \alpha_{(z,\tau_i x)}(\alpha_{(\tau_i x,x)}(a_{\tau_i}(x)))c(x,\tau_i x,z)f(\tau_i x,z)\\
&= \sum_{i=1}^{n_k} \alpha_{(z,\tau_i x)}(\alpha_{(\tau_i x,x)}(\chi_{E_{\tau_i}}(x)a_{\tau_i}(x)))c(x,\tau_i x,z)f(\tau_i x,z)\\
&= \begin{cases} \alpha_{(z,\tau_i x)}(\alpha_{(\tau_i x,x)}(\chi_{E_{\tau_i}} a_{\tau_i}(x)))c(x,\tau_i x,z)f(\tau_i x,z) & \\ \qquad \text{if there exists } i \le n_k \text{ such that } x \in E_{\tau_i} & \\ 0 \qquad \text{otherwise} & \end{cases} \qquad (12.4)
\end{aligned}$$

As $k \to \infty$, the last expression converges to:

$$\alpha_{(z,\tau x)}(\alpha_{(\tau x,x)}(a_\tau(x)))c(x,\tau x,z)f(\tau x,z)$$

where $\tau \in T$ is such that $x \in E_\tau$, or to

$$\begin{aligned}
&\alpha_{(z,\phi_V x)}(\alpha_{(\phi_V x,x)}(a_\tau(x)))c(x,\phi_V x,z)f(\phi_V x,z)\\
&= \textstyle\sum_{y\sim x} \alpha_{(z,y)}(V(x,y))c(x,y,z)f(y,z)
\end{aligned}$$

where

$$V(x,y) = \begin{cases} \alpha_{(\phi_V x,x)}(a_\tau(x)) & \text{if } x \in E_\tau,\ y = \phi_V x \\ 0 & \text{otherwise} \end{cases}$$

On the other hand, the expression (12.4) converges to $(Vf)(x,z)$. Thus, the theorem is proved in this case.

2. Let $V = \beta(I(a))$, $a \in M$. By Theorem 12.28:

$$\begin{aligned}
&(Vf)(x,z) = \alpha_{(z,x)}(a(x))f(x,z)\\
&= \textstyle\sum_{y\sim z} \alpha_{(z,y)}(V(x,y))c(x,y,z)f(y,z)
\end{aligned}$$

where $V(x,y) = \begin{cases} a(x) & \text{if } x = y \\ 0 & \text{if } x \ne y \end{cases}$. (Note: $c(x,x,z) = 1_{H(z)}$.)

3. Let $V = \beta(I(a))V_1$ where $V_1 \in \beta(N)$.

$$\begin{aligned}
&(Vf)(x,z) = (\beta(I(a))V_1 f)(x,z) = \alpha_{(z,x)}(a(x))(V_1 f)(x,z)\\
&= \alpha_{(z,x)}(a(x))\alpha_{(z,x)}(a_\tau(x))c(x,\phi_{V_1}x,z)f(\phi_{V_1}x,z)\\
&= \alpha_{(z,x)}(a(x)a_\tau(x))c(x,\phi_{V_1}x,z)f(\phi_{V_1}x,z)\\
&= \alpha_{(z,\phi_{V_1}x)}(\alpha_{(\phi_{V_1}x,z)}(a(x)a_\tau(x)))c(x,\phi_{V_1}x,z)f(\phi_{V_1}x,z)\\
&= \textstyle\sum_{y\sim x} \alpha_{(z,y)}(V(x,y))c(x,y,z)f(y,z)
\end{aligned}$$

where $V(x,y) = \begin{cases} \alpha_{(\phi_{V_1}x,x)}(a(x)a_\tau(x)) & \text{if } y = \phi_{V_1}x \\ 0 & \text{otherwise} \end{cases}$.

4. Let $V = \sum_{i=1}^{n} \beta(I(a_i))V_i$, where $a_i \in M$, $V_i \in \beta(N)$, $i = 1, \ldots, n$. Then:

$$\begin{aligned}(Vf)(x,z) &= \textstyle\sum_{i=1}^{n} (\beta(I(a_i))V_i f)(x,z)\\ &= \textstyle\sum_{i=1}^{n} \sum_{y\sim x} \alpha_{(z,y)}((\beta(I(a_i))V_i)(x,y))c(x,y,z)f(y,z)\\ &= \textstyle\sum_{y\sim x} \alpha_{(z,y)}(\sum_{i=1}^{n}(\beta(I(a_i))V_i)(x,y))c(x,y,z)f(y,z)\end{aligned}$$

5. Every $V_i \in \beta(N)$ normalises $\beta(I(M))$, so every element $V \in \tilde{M}_0$ can be represented in the form $V = \sum_{i=1}^{n} \beta(I(a_i))V_i$. Thus, the theorem is proved for every $V \in \tilde{M}_0$.

□

REMARK. Here for $V \in \beta(N)$: $\phi_V = \phi_{\beta^{-1}(V)}$.

NOTATION. Let $\mathcal{D}(\tau)$ mean the domain and $\mathcal{R}(\tau)$ mean the range of a partial Borel automorphism τ of X.

Corollary 12.33 *For every $\tau \in T$ there exists $W_\tau \in \tilde{M}_0$ such that*

$$W_\tau(x,y) = \begin{cases} 1_{H(y)} & \textit{if } y = \tau x \\ 0 & \textit{otherwise} \end{cases}$$

so that $(W_\tau f)(x,z) = \chi_{\mathcal{D}(\tau)}(x)c(x,\tau x,z)f(\tau x,z)$.

PROOF. Take $V = U_\tau$ in part (1) of the proof of Theorem 12.32. Then $E_\tau = \mathcal{D}(\tau)$, $V_\tau = \beta(I(\chi_{\mathcal{D}(\tau)}))U_\tau$, $a_\tau = \chi_{\mathcal{D}(\tau)}$ and

$$V_\tau(x,y) = \begin{cases} 1_{H(y)} & \text{if } y = \tau x \\ 0 & \text{otherwise} \end{cases}$$

Take $W_\tau = V_\tau$.

□

Corollary 12.34 *For every $b \in M$, $\tau \in T$ there exists $V \in \tilde{M}_0$ such that*

$$V(x,y) = \begin{cases} b(y) & \textit{if } y = \tau x \\ 0 & \textit{otherwise} \end{cases}$$

PROOF. Let $V = \beta(I(a))W_\tau$, $a \in M$, a_τ be as in part (1) of the proof of Theorem 12.32.

$$V(x,y) = \begin{cases} \alpha_{(y,x)}(a(x)a_\tau(x)) & \text{if } y = \tau x \\ 0 & \text{otherwise} \end{cases} = \begin{cases} \alpha_{(y,x)}(a(x)) & \text{if } y = \tau x \\ 0 & \text{otherwise} \end{cases}$$

because in this case $a_\tau = \chi_{\mathcal{D}(x)}$. Now, let $a \in M$ be such that $b(\tau x) = \alpha_{(x,\tau x)}(b(\tau x))$, i. e. $a(x) = \alpha_{(x,\tau x)}(b(\tau x))$.

□

REMARK. The operators $V(x,y)$ are uniquely defined by V (up to a null set). To see this, it is enough to take $f \in \widetilde{\widetilde{H}}$ with $\operatorname{supp} f \subset \Gamma(\tau)$, $\tau \in \Sigma$. Then the sum turns to a single summand.

Theorem 12.35 (See Theorem 5.2(5.2).) *For $T_1, T_2 \in \tilde{M}_0$:*

$$(T_1T_2)(x,z) = \sum_{y\sim x} \alpha_{(z,y)}(T_1(x,y))c(x,y,z)T_2(y,z)c(y,z,z)c(x,z,z)^{-1}$$

PROOF. Let $f \in \widetilde{\widetilde{H}}$, $\operatorname{supp} f \subset \Delta$.

$$\begin{aligned}(T_1T_2)(x,z) &= (T_1(T_2(f))(x,z) \\ &= \textstyle\sum_y \alpha_{(z,y)}(T_1(x,y))c(x,y,z)(T_2f)(y,z) \\ &= \textstyle\sum_y \alpha_{(z,y)}(T_1(x,y))c(x,y,z) \\ &\qquad \textstyle\sum_w \alpha_{(z,w)}(T_2(y,w))c(y,w,z)f(w,z) \\ &= \textstyle\sum_y \alpha_{(z,y)}(T_1(x,y))c(x,y,z)T_2(y,z)c(y,z,z)f(z,z)\end{aligned}$$

On the other hand:

$$\begin{aligned}(T_1T_2f)(x,z) &= \textstyle\sum_w ((T_1T_2)(x,w))c(x,w,z)f(w,z) \\ &= (T_1T_2)(x,z)c(x,z,z)f(z,z)\end{aligned}$$

This holds for every $f \in \widetilde{\widetilde{H}}$ with $\operatorname{supp} f \subset \Delta$. Thus,

$$(T_1T_2)(x,z) = \sum_{y} \alpha_{(z,y)}(T_1(x,y))c(x,y,z)T_2(y,z)c(y,z,z)c(x,z,z)^{-1}$$

□

Corollary 12.36 *The function $c : R^2 \to \bigcup_{z\in X} B(H(z))$ defined above satisfies the properties (1)–(6) of c which are formulated at the end of Section 2.*

PROOF. The properties (1) and (6) are obvious, the property (5) follows from the equality (12.2) before Theorem 12.32. Finally, the properties (2), (3) and (4) can be proved by developing the equalities $\langle Tf, g\rangle = \langle f, T^*g\rangle$ and $(T_1T_2)f = T_1(T_2f)$ using Theorems 12.32 and 12.35.

□

Lemma 12.37 *For every $\sigma \in \Sigma$ the operator $\pi(\sigma)$ belongs to $\beta(\tilde{M})$, where*

$$(\pi(\sigma)f)(x,y) = c(x,\sigma x,y)f(\sigma x,y), \qquad f \in \widetilde{\widetilde{H}}$$

PROOF. (Like in part (1) of the proof of Theorem 12.32.) Let $\sigma \in \Sigma$. For every $\tau \in T$ let $E_\tau = \{x \in X \mid \sigma x = \tau x\}$. Then $\bigcup_{\tau \in T} E_\tau = X$ is a disjoint union, so $\sum \chi_{E_\tau} = 1_H$, hence $\pi(\sigma) = \sum_{\tau \in T} \beta(I(\chi_{E_\tau}))\pi(\sigma)$ where the sum is a strong limit of a sequence of partial sums (T is countable !).

The algebra $\beta(\tilde{M})$ is the closure of $\tilde{M}_0$, hence it is enough to prove that $\beta(I(\chi_{E_\tau}))\pi(\sigma) \in \tilde{M}_0$ for every $\tau \in T$. Let $f \in \tilde{\tilde{H}}$.

$$
\begin{aligned}
&(\beta(I(\chi_{E_\tau}))\pi(\sigma)f)(x,y) = \chi_{E_\tau}(x)(\pi(\sigma)f)(x,y) = \chi_{E_\tau}(x)c(x,\sigma x,y)f(\sigma x,y) \\
&= \chi_{E_\tau}(x)c(x,\tau x,y)f(\tau x,y) = (W_\tau f)(x,y)
\end{aligned}
$$

Thus, $\beta(I(\chi_{E_\tau}))\pi(\sigma) = W_\tau \in \tilde{M}_0$.

□

Corollary 12.38 *$\beta(\tilde{M})$ is generated by $\beta(I(M))$ and all of $\pi(\sigma)$ for $\sigma \in \Sigma$.*

PROOF. By Lemma 12.37, $\beta(\tilde{M})$ contains all of $\pi(\tau)$. By the definition., $\beta(\tilde{M})$ contains $\beta(I(M))$. Furthermore, $\tilde{M}$ is generated by N, so $\beta(\tilde{M})$ is generated by $\beta(N)$. Let $V \in \beta(N)$, $f \in \tilde{\tilde{H}}$. By part (1) of the proof of Theorem 12.32,

$$
\begin{aligned}
&(Vf)(x,z) = \alpha_{(z,\phi_V x)}(V(x,\phi_V x))c(x,\phi_V x,z)f(\phi_V x,z) \\
&= \beta(I(a))\pi(\phi_V)f(x,z)
\end{aligned}
$$

where $a \in M$ is such that $a(y) = V(\phi_V^{-1} y, y)$.

Thus, every $V \in \beta(N)$ is a combination of elements of the form $\beta(I(a))$ with $a \in M$ and of the form $\pi(\tau)$ for $\tau \in \Sigma$. Hence, $\beta(\tilde{M})$ is generated by $\beta(I(M))$ and $\pi(\tau)$ with $\tau \in \Sigma$.

□

Thus, $\beta(\tilde{M})$ is just the crossed product of M with R. The structure theorem (Theorem 12.2) is proved.

□

CHAPTER 13

ISOMORPHISMS OF CROSSED PRODUCTS

13.1 $I(M)$-isomorphisms of crossed products

Let X, M, R etc. be as above.

Definition 13.1 *The map $c : R^2 \to \bigcup_{z\in X} B(H(z))$ satisfying the properties 1–6 of c at the end of Section 2 will be called a* **"2-cocycle"**.

The collection $\alpha = \{\alpha_{(x,y)}\}_{(x,y)\in R}$ satisfying the properties 1–3 of α at the end of Section 2 will be called the **set of correspondence maps**.

Two "2-cocycles" c_1 and c_2 will be said to be **"cohomologous"** *(with respect to a given couple (α^1, α^2) of sets of correspondence maps) if there exist maps $V : R \to \bigcup_{y\in X} B(H(y))$, $Q : R \to \bigcup_{y\in X} B(H(y))$ such that:*

1. *for every $(x, y) \in R$, $V(x, y)$ is a unitary operator in $U^2_{(y,x)} M'(x) U^1_{(x,y)}$; (Here $U^i_{(x,y)}$ is the unitary operator corresponding to $\alpha^i_{(x,y)}$ as in [6, Corollary 2.5.32].)*

2. *for every $(x, y) \in R$, $Q((x, y)$ is a unitary operator in $M(y)$; for $x \in X$: $Q(x, x) = 1_{H(x)}$;*

3. *for every $\sigma \in \Sigma$, the maps $x \mapsto V(x, \sigma x)$ and $x \mapsto Q(x, \sigma x)$ are measurable;*

4. *for a. e. $x \sim y \sim z$:*

$$c_2(x, y, z) = V(x, z)\alpha^1_{(z,y)}(Q(x, y))c_1(x, y, z)V(y, z)^*$$

Let $\tilde{M}_1$ and $\tilde{M}_2$ be the crossed products of the same von Neumann algebra $M = \int_X M(x)d\mu(x)$ by the same equivalence relation $R \subset X \times X$, but with respect to different sets of correspondence maps α^1 and α^2 an different "cocycles" c_1 and c_2 respectively.

We will denote all objects associated with $\tilde{M}_1$ or $\tilde{M}_2$ by a subscript or superscript 1 or 2 respectively.

Definition 13.2 *Two such crossed products $\tilde{M}_1$ and $\tilde{M}_2$ will be called* ***$I(M)$*-isomorphic** *if there exists an isomorhism $\kappa : \tilde{M}_1 \to \tilde{M}_2$ such that $\kappa(I_1(a)) = I_2(a)$ for every $a \in M$.*

Theorem 13.3 *Let $\tilde{M}_1$ and $\tilde{M}_2$ be two different crossed products of M with R as above. $\tilde{M}_1$ and $\tilde{M}_2$ are $I(M)$-isomorphic if and only if c_1 and c_2 are "cohomologous" with respect to the couple (α^1, α^2).*

PROOF. Suppose first that $\tilde{M}_1$ is $I(M)$-isomorphic to $\tilde{M}_2$. Let κ be the corresponding isomorphism. Let $\tilde{M} = \tilde{M}_1 \oplus \tilde{M}_2$, $\tilde{H} = \tilde{H}_1 \oplus \tilde{H}_2$. Then the vector $\tilde{\phi}_0 = \tilde{\phi}_0^1 \oplus \tilde{\phi}_0^2$ is cyclic and separating for $\tilde{M}$. Let $\tilde{S}$, $\tilde{F}$, $\tilde{\Delta}$, $\tilde{J}$ be the operators from the Tomita-Takesaki theory corresponding to $\tilde{M}$ and $\tilde{\phi}_0$. Then, by the definitions, $\tilde{S} = \tilde{S}_1 \oplus \tilde{S}_2$, $\tilde{F} = \tilde{F}_1 \oplus \tilde{F}_2$, and so $\tilde{\Delta} = \tilde{\Delta}_1 \oplus \tilde{\Delta}_2$ and $\tilde{J} = \tilde{J}_1 \oplus \tilde{J}_2$.

Let $\tilde{\kappa} = \kappa \oplus \kappa^{-1}$. Then $\tilde{\kappa}$ is an isomorphism of $\tilde{M} = \tilde{M}_1 \oplus \tilde{M}_2$ onto $\tilde{M}_2 \oplus \tilde{M}_1$. Let θ be the isomorphism of $\tilde{M}_2 \oplus \tilde{M}_1$ onto $\tilde{M} = \tilde{M}_1 \oplus \tilde{M}_2$ defined by $\theta(a \oplus b) = b \oplus a$ for $a \in \tilde{M}_2$, $b \in \tilde{M}_1$. Then $\theta \circ \tilde{\kappa}$ is an automorphism of $\tilde{M}$. Since $\tilde{M}$ possesses a cyclic and separating vector, $\theta \circ \tilde{\kappa} = \mathrm{Ad}V$ fore some unitary operator V on $\tilde{H}$ commuting with $\tilde{J}$.

Let $P_1 = 1_{\tilde{H}_1} \oplus 0_{\tilde{H}_2}$, $P_2 = 0_{\tilde{H}_1} \oplus 1_{\tilde{H}_2}$. We have:

$$\begin{aligned}
&\theta \circ \tilde{\kappa}(P_1) = \theta(\tilde{\kappa}(P_1)) = \theta((\kappa \oplus \kappa^{-1})(1_{\tilde{H}_1} \oplus 0_{\tilde{H}_2})) \\
&= \theta(1_{\tilde{H}_2} \oplus 0_{\tilde{H}_1}) = 0_{\tilde{H}_1} \oplus 1_{\tilde{H}_2} = P_2; \\
&\theta \circ \tilde{\kappa}(P_2) = \theta(\tilde{\kappa}(P_2)) = \theta((\kappa \oplus \kappa^{-1})(0_{\tilde{H}_1} \oplus 1_{\tilde{H}_2})) \\
&= \theta(0_{\tilde{H}_2} \oplus 1_{\tilde{H}_1}) = 1_{\tilde{H}_1} \oplus 0_{\tilde{H}_2} = P_1
\end{aligned}$$

Thus, $VP_1V^* = P_2$, $VP_2V^* = P_1$, hence $VP_1 = P_2V$ and $VP_2 = P_1V$. Thus, for $f \in P_1\tilde{H} = \tilde{H}_1 \oplus 0$: $Vf = VP_1f = P_2Vf$, so $Vf \in P_2\tilde{H} = 0 \oplus \tilde{H}_2$.

Similarly, for $f \in P_2\tilde{H} = 0 \oplus \tilde{H}_2$: $Vf = VP_2f = P_1Vf$, so $Vf \in P_1\tilde{H} = \tilde{H}_1 \oplus 0$.

Let $V_1 : \tilde{H}_1 \to \tilde{H}_2$, $V_2 : \tilde{H}_2 \to \tilde{H}_1$ be defined by:

$$\begin{aligned}
V(f \oplus 0) &= 0 \oplus V_1 f, \quad f \in \tilde{H}_1 \\
V(0 \oplus g) &= V_2 g \oplus 0, \quad g \in \tilde{H}_2
\end{aligned}$$

Then for $f \in \tilde{H}_1$, $g \in \tilde{H}_2$: $V(f \oplus g) = V_2g \oplus V_1f$. So, $V^{-1}(V_2g \oplus V_1f) = f \oplus g$, or:

$$V^{-1}(h \oplus k) = V_1^{-1}k \oplus V_2^{-1}h, \qquad h \in \tilde{H}_1, \quad g \in \tilde{H}_2$$

Thus, for $a \oplus b \in \tilde{M}$ and for $f \oplus g \in \tilde{H}$:

$$\begin{aligned}
&(\theta \circ \tilde{\kappa})(a \oplus b) = V(a \oplus b)V^{-1}(f \oplus g) = V(a \oplus b)(V_1^{-1}g \oplus V_2^{-1}f) \\
&= V(aV_1^{-1}g \oplus bV_2^{-1}f) = V_2bV_2^{-1}f \oplus V_1aV_1^{-1}g = (V_2bV_2^{-1} \oplus V_1aV_1^{-1})(f \oplus g)
\end{aligned}$$

Hence $\theta \circ \tilde{\kappa}(a \oplus b) = V_2bV_2^{-1} \oplus V_1aV_1^{-1}$, and $\tilde{\kappa}(a \oplus b) = V_1aV_1^{-1} \oplus V_2bV_2^{-1}$. But $\tilde{\kappa} = \kappa \oplus \kappa^{-1}$. Thus, $\kappa = \mathrm{Ad}\, V_1$, $\kappa^{-1} = \mathrm{Ad}\, V_2$. Furthermore, V commutes with $\tilde{J}$, so for $f \oplus g \in \tilde{H} = \tilde{H}_1 \oplus \tilde{H}_2$:

$$\begin{aligned}
&V_2\tilde{J}_2g \oplus V_1\tilde{J}_1f = V(\tilde{J}_1f \oplus \tilde{J}_2g) = V(\tilde{J}_1 \oplus \tilde{J}_2)(f \oplus g) \\
&= V\tilde{J}(f \oplus g) = \tilde{J}V(f \oplus g) = \tilde{J}(V_2g \oplus V_1f) = \tilde{J}_1V_2g \oplus \tilde{J}_2V_1f
\end{aligned}$$

Thus, $V_2\tilde{J}_2 = \tilde{J}_1V_2$ and $V_1\tilde{J}_1 = \tilde{J}_2V_1$.

For a Borel subset $E \subseteq X$ we define:

$$\begin{aligned}
(P_E^i f)(x, y) &= \chi_E(x)f(x, y), \quad f \in \tilde{H}_i, \quad i = 1, 2 \\
(Q_E^i f)(x, y) &= \chi_E(y)f(x, y), \quad f \in \tilde{H}_i, \quad i = 1, 2
\end{aligned}$$

We have: $P_E^i = I_i(\chi_E)$ for $i = 1, 2$, so $V_1 P_E^1 V_1^{-1} = P_E^2$ and

$$V_1 P_E^1 = P_E^2 V_1 \tag{13.1}$$

Furthermore, $Q_E^i = \tilde{J}_i P_E^i \tilde{J}_i$ for $i = 1, 2$, so:

$$\begin{aligned} V_1 Q_E^1 &= V_1 \tilde{J}_1 P_E^1 \tilde{J}_1 = \tilde{J}_2 V_1 P_E^1 \tilde{J}_1 \\ &= \tilde{J}_2 P_E^2 V_1 \tilde{J}_1 = \tilde{J}_2 P_E^2 \tilde{J}_2 V_1 = Q_E^2 V_1 \end{aligned}$$

Thus,

$$V_1 Q_E^1 = Q_E^2 V_1 \tag{13.2}$$

Hence, for every Borel subsets $E, F \subset X$:

$$V_1 P_E^1 Q_F^1 = P_E^2 Q_F^2 V_1$$

Hence, for every Borel subset $K \subset R$:

$$V_1 \chi_K^1 = \chi_K^2 V_1$$

where $\chi_K^i \in B(\tilde{H}_i)$ and

$$(\chi_K^i f)(x, y) = \chi_K(x, y) f(x, y), \quad f \in \tilde{H}_i \quad i = 1, 2$$

By [10, Ch. II, §2, n. 5, Théorème 1], V_1 is a decomposable operator on $\tilde{H}_1 = \int_R H(x, y) d\nu(x, y)$, i. e. for $f \in \tilde{H}_1$:

$$(V_1 f)(x, y) = V_1(x, y) f(x, y)$$

where $V_1(x, y) \in B(H(x, y))$, $H(x, y) \equiv H(y)$, $(x, y) \in R$. Since V_1 is unitary, almost every $V(x, y)$ is unitary. Furthermore, for every $a \in M$:

$$I_2(a) = \kappa(I_1(a)) = V_1 I_1(a) V_1^{-1}$$

In other words, $V_1 I_1(a) = I_2(a) V_1$.

Let $f \in \tilde{H}_1$. Then:

$$\begin{aligned} &(V_1 I_1(a) f)(x, y) = V_1(x, y)(I_1(a) f)(x, y) \\ &= V_1(x, y) \alpha^1_{(y,x)}(a(x)) f(x, y); \\ &(I_2(a) V_1 f)(x, y) = \alpha^2_{(y,x)}(a(x))(V_1 f)(x, y) \\ &= \alpha^2_{(y,x)}(a(x)) V_1(x, y) f(x, y) \end{aligned}$$

Thus, for a. e. $(x, y) \in R$:

$$\begin{aligned} V_1(x, y) \alpha^1_{(y,x)}(a(x)) &= \alpha^2_{(y,x)}(a(x)) V_1(x, y); \\ V_1(x, y) U^1_{(y,x)} a(x) U^1_{(x,y)} &= U^2_{(y,x)} a(x) U^2_{(x,y)} V_1(x, y); \\ U^2_{(x,y)} V_1(x, y) U^1_{(y,x)} a(x) &= a(x) U^2_{(x,y)} V_1(x, y) U^1_{(y,x)} \end{aligned}$$

This holds for every $a \in M$. Thus, for a. e. $(x, y) \in R$:

$$U^2_{(x,y)} V_1(x, y) U^1_{(y,x)} \in M'(x)$$

In other words:

$$V_1(x,y) \in U^2_{(y,x)} M'(x) U^1_{(x,y)}$$

Let $T \in \tilde{M}_1$, $S = \kappa(T) = V_1 T V_1^{-1} \in \tilde{M}_2$, $f \in \tilde{H}_2$. Then

$$(Sf)(x,y) = \\ = \sum_z V_1(x,y)\alpha^1_{(y,z)}(T(x,z))c_1(x,z,y)V_1^{-1}(z,y)f(z,y)$$

On the other hand,

$$(Sf)(x,y) = \sum_z \alpha^2_{(y,z)} c_2(x,z,y) f(z,y)$$

This is true for every $f \in \tilde{H}_2$, so

$$V_1(x,y)\alpha^1_{(y,z)}(T(x,z))c_1(x,z,y)V_1^{-1}(z,y) \\ = \alpha^2_{(y,z)}(S(x,z))c_2(x,z,y)$$

Let T_2 be a partition of R_2 as in Lemma 12.31. For every $\tau \in T_2$ choose $S_\tau \in \tilde{M}_2$ such that for $(x,z) \in \Gamma(\tau)$: $S_\tau(x,z) = 1_H(z)$. Let $T_\tau \in \tilde{M}_1$ correspond to this S_τ (i. e. $T_\tau = V_1^{-1} S V_1$). Let us denote $T_\tau(x,z)$ for $(x,z) \in \Gamma(\tau)$ by $Q(x,z)$.

Then for $(x,z) \in \Gamma(\tau)$, $y \sim x$:

$$V_1(x,y)\alpha^1_{(y,z)}(Q(x,z))c_1(x,z,y)V_1^{-1}(z,y) = c_2(x,z,y)$$

This holds for every $\tau \in T_2$ so for almost every $x \sim y \sim z$.

We have:

1. $Q(x,z) \in M(z)$;
2. $Q(x,x) = 1_H(x)$ (in this case $\tau = \mathrm{id}_X$ and $\Gamma(\tau) = \triangle$);
3. $Q(x,z) = \alpha^1_{(z,y)}(V(x,y)^{-1}c_2(x,z,y)V_1(z,y)c_1(x,z,y)^{-1})$ is unitary.

Thus, c_1 and c_2 are "cohomologous".

Conversely, let c_2 be "cohomologous" to c_1, i. e. $V(x,y) = U^2_{(y,x)} v(y,x) U^1_{(x,y)}$ where $v(y,x) \in M'(x)$ and

$$c_2(x,z,y) = V(x,y)\alpha^1_{(y,z)}(Q(x,z))c_1(x,z,y)V(z,y)^*$$

where V and Q are as above.

Let $W \in B(\tilde{H}_1, \tilde{H}_2)$ be defined by

$$(Wf)(x,y) = V(x,y)f(x,y), \qquad f \in \tilde{H}_1$$

Then for $g \in \tilde{H}_2$: $(W^*g)(x,y) = V(x,y)^*g(x,y)$. Let $\rho = \mathrm{Ad}\,(W^*)$. Then ρ is an isomorphism of $B(\tilde{H}_2)$ onto $B(\tilde{H}_1)$.

Let $T \in \tilde{M}_2$. The straightforward calculation shows that the operator $\rho(T) = W^*TW$ acts as follows:

$$(\rho(T)f)(x,y) \\ = \sum_z \alpha^1_{(y,z)}(\rho(T)(x,z))c_1(x,z,y)f(z,y)$$

where

$$\rho(T)(x,z) = \alpha^1_{(z,x)} \circ \alpha^2_{(x,z)}(T(x,z))Q(x,z) \in M(z)$$

To prove that $\rho(T) \in \tilde{M}_1$ it is enough to check that $\rho(T)$ commutes with $I'_1(a)$ for $a \in M'$ and with $\pi'_1(\tau)$ for $\tau \in \Sigma$. It can be verified immediately.

A similar calculation shows that if $\rho(T) \in \tilde{M}_1$ then $T \in \tilde{M}_2$. Thus $\rho(\tilde{M}_2) = \tilde{M}_1$.

Let $T = I_2(a)$, $a \in M$. Then $T(x,y) = \begin{cases} a(y) & \text{if } x = y \\ 0 & \text{otherwise} \end{cases}$. Then

$$\rho(T)(x,y) = \alpha^1_{(y,x)} \circ \alpha^2_{(x,y)}(T(x,y))Q(x,y)$$

$$= \begin{cases} a(y) & \text{if } x = y \\ 0 & \text{otherwise} \end{cases}$$

i. e. $\rho(T) = I_1(a)$.

Thus, for $a \in M$: $\rho(I_2(a)) = I_1(a)$. Hence, $\tilde{M}_2$ is isomorphic to $\tilde{M}_1$.

□

Corollary 13.4 *In the structure theorem (Theorem 12.2), the "cocycle" c is unique up to "cohomology".*

13.2 I-isomorphisms of crossed products

Let $i = 1$ or $i = 2$. Let M_i, X_i, R_i, $\tilde{M}_i$ etc. be as above.

Lemma 13.5 *Let $\tilde{M} = M \bowtie R$, V be a unitary operator in $\tilde{M}$ which normalises $I(A)$ (i. e. $VI(A)V^* = I(A)$). Let τ be the corresponding Borel automorphism of X (i. e. for $a \in A$: $VI(a)V^* = I(a \circ \tau^{-1})$). Then up to a null set: $\Gamma(\tau) \subset R$.*

Proof. The operator V normalises (by the trivial way) the algebra $B = \tilde{J}I(A)\tilde{J}$. Hence, it normalises the algebra $C \cong L^\infty(R,\nu)$, generated by $I(A)$ and B. Let $\tilde{\tau}$ be the corresponding Borel automorphism of R.

By the same calculation as in the proof of Lemma 12.9 we can see that $\tilde{\tau}$ is of the form $\tilde{\tau}(x,y) = (\tau x, y)$, $(x,y) \in R$. Thus, for a. e. $(x,y) \in R$: $(\tau x, y) \in R$. In particular, $(\tau x, x) \in R$. Hence $\Gamma(\tau) \subset R$.

□

Lemma 13.6 *Let κ be an isomorphism of $\tilde{M}_1$ onto $\tilde{M}_2$, mapping $I_1(A_1)$ onto $I_2(A_2)$. Let $\kappa|I_1(A_1) : I_1(A_1) \to I_2(A_2)$ be correspond to a Borel isomorphism $\tau : X_1 \to X_2$, i. e. for $a \in A_1$:*

$$\kappa(I_1(a)) = I_2(a \circ \tau^{-1})$$

Then $\tau^{(2)}(R_1) = R_2$.

PROOF. Let $\sigma \in \Sigma_1$, $V = \kappa(\pi(\sigma))$. Then for $a \in A_1$:

$$\begin{aligned} VI_2(a)V^* &= \kappa(\pi(\sigma)) \cdot \kappa(I_1(a \circ \tau^{-1})) \cdot \kappa(\pi(\sigma)^{-1}) \\ &= \kappa(\pi(\sigma) I_1(a \circ \tau^{-1}) \pi(\sigma)^{-1}) = \kappa(I_1(a \circ \tau \circ \sigma)) = I_2(a \circ \tau \circ \sigma \circ \tau^{-1}) \end{aligned}$$

Thus, the isomorphism $\operatorname{Ad} V$ corresponds to the Borel isomorphism $\tau \circ \sigma^{-1} \circ \tau^{-1}$. By Lemma 13.5, $\Gamma(\tau \circ \sigma^{-1} \circ \tau^{-1}) \subset R_2$. This is true for every $\sigma \in \Sigma_1$. Hence $\tau^{(2)}(R_1) \subseteq R_2$.

Applying all these considerations to κ^{-1} and τ^{-1} we get: $(\tau^{-1})^{(2)}(R_2) \subseteq R_1$. Hence $\tau^{(2)}(R_1) = R_2$.

□

Lemma 13.7 (See [16, Proposition 4.1].) *For $i = 1$ and $i = 2$, let X_i be a locally compact space, μ_i be a Borel measure on X_i, $H_i = \int_{X_i} H_i(x) d\mu_i(x)$ be a Hilbert integral, Z_i be the algebra of diagonalisable operators on H_i.*

Let τ be a Borel map from X_1 onto X_2 such that $\mu_2 \circ \tau \sim \mu_1$. Let $r(x) = (d(\mu_2 \circ \tau)/d\mu_1)(x)$.

Let U be a unitary operator from H_1 onto H_2 such that for every $f \in L^\infty(X_1) \cong Z_1$: $UfU^{-1} = f \circ \tau^{-1}$.

Then there exist unitary operators $U(x)$ from $H_1(x)$ onto $H_2(\tau x)$, $x \in X_1$, such that for every $\xi \in H_1$, a. e. $y \in X_2$:

$$(U\xi)(y) = r(\tau^{-1}y)^{-1/2} U(\tau^{-1}y) \xi(\tau^{-1}y)$$

PROOF. We can assume that $X_1 \cap X_2 = \emptyset$. Let $X = X_1 \cup X_2$, and for a Borel set $E \subset X$ let $\mu(E) = \mu_1(E \cap X_1) + \mu_2(E \cap X_2)$. Then μ is a Borel measure on X. Let $H = H_1 \oplus H_2$. Then $H = \int_X H(x) d\mu(x)$ where
$H(x) = \begin{cases} H_1(x) & \text{if } x \in X_1 \\ H_2(x) & \text{if } x \in X_2 \end{cases}$.

Let $\bar{U} = U \oplus U^*$. Then $\bar{U}$ acts from $H = H_1 \oplus H_2$ onto $H_2 \oplus H_1$.

Let $\theta : H_2 \oplus H_1 \to H_1 \oplus H_2$ be defined as follows: $\theta((f_2, f_1)) = (f_1, f_2)$, $f_i \in H_i$. Let $V = \theta \circ \bar{U}$. Then V is a unitary operator on H.

Let $\bar{\tau}(x) = \begin{cases} \tau(x) & \text{if } x \in X_1 \\ \tau^{-1}(x) & \text{if } x \in X_2 \end{cases}$. Then $\bar{\tau}$ is a Borel map on X, and $\bar{\tau}^{-1} = \bar{\tau}$.

For a Borel subset $E \subseteq X$:

$$\begin{aligned} &\int_E d\mu(\bar{\tau}x) = \int_{E \cap X_1} d\mu(\bar{\tau}x) + \int_{E \cap X_2} d\mu(\bar{\tau}x) \\ &= \int_{E \cap X_1} d\mu(\tau x) + \int_{E \cap X_2} d\mu(\tau^{-1}x) \\ &= \int_{E \cap X_1} d\mu_2(\tau x) + \int_{E \cap X_2} d\mu_1(\tau^{-1}x) \\ &= \int_{E \cap X_1} r(x) d\mu_1(x) + \int_{E \cap X_2} r(\tau^{-1}x)^{-1} d\mu_2(x) = \int_E \bar{r}(x) d\mu(x) \end{aligned}$$

where $\bar{r}(x) = \begin{cases} r(x) & \text{if } x \in X_1 \\ r(\tau^{-1}x)^{-1} & \text{if } x \in X_2 \end{cases}$.

Now, let $f \in L^\infty(X, \mu)$. Then $f = f_1 + f_2$, where $f_i \in L^\infty(X_i, \mu_i) \cong Z_i$, $i = 1, 2$. Let $\xi \in H = H_1 \oplus H_2$, $\xi = (\xi_1, \xi_2)$.

$$\begin{aligned} VfV^*\xi &= \theta \bar{U} f \bar{U}^* \theta^* \xi \\ &= \theta \bar{U} f \bar{U}^* (\xi_2, \xi_1) = \theta \bar{U} f (U^* \xi_2, U \xi_1) \end{aligned}$$

$$= \theta\bar{U}(f_1U^*\xi_2, f_2U\xi_1) = \theta(Uf_1U^*\xi_2, U^*f_2U\xi_1)$$
$$= (U^*f_2U\xi_1, Uf_1U^*\xi_2) = ((f_2\circ\tau)\xi_1, (f_1\circ\tau^{-1})\xi_2)$$
$$= (f\circ\bar{\tau})\xi = (f\circ\bar{\tau}^{-1})\xi$$

By [16, Proposition 4.1], there exist unitary operators $V(x)$ from $H(x)$ onto $H(\bar{\tau}x)$ such that for $\xi \in H$, a. e. $x \in X$:

$$(V\xi)(x) = \bar{r}(\bar{\tau}^{-1}x)V(\bar{\tau}^{-1}x)\xi(\bar{\tau}^{-1}x)$$

Take $x \in X_2$, $\xi \in H_1$ and get the required.

□

Definition 13.8 *Two crossed products as above $\tilde{M}_1$ and $\tilde{M}_2$ will be called* ***I*-isomorphic**, *if there exists a *-isomorhism κ from $\tilde{M}_1$ onto $\tilde{M}_2$ such that*

$$\kappa(I_1(M_1)) = I_2(M_2)) \quad \text{and} \quad \kappa(I_1(A_1)) = I_2(A_2))$$

Theorem 13.9 (See [15, Theorem 2].) *Two crossed products $\tilde{M}_1$ and $\tilde{M}_2$ are I-isomorphic if and only if there exist:*

1. *a Borel isomorphism $\tau : X_1 \to X_2$ such that $\mu_1 \sim \mu_2 \circ \tau$ and $\tau^{(2)}(R_1) = R_2$;*
2. *a map $V : R_2 \to \bigcup_{y\in X_2} B(H_1(\tau^{-1}y), H_2(y))$ such that for a. e. $(x, y) \in R_2$: V(x,y) is a unitary operator acting from $H_2(y)$ onto $H_1(\tau^{-1}y)$ and such that the isomorphism*

$$\mathrm{Ad}\left(U^1_{(\tau^{-1}x,\tau^{-1}y)}V(x,y)U^2_{(y,x)}\right) : M_2(x) \to M_1(\tau^{-1}x)$$

   does not depend on y;

3. *a map $Q : R_2 \to \bigcup_{y\in X_2} M_2(y)$ such that for a. e. $(x, y) \in R_2$: $Q(x, y)$ is a unitary operator in $M_2(y)$, for a. e. $x \in X_2$: $Q(x, x) = 1_{H(x)}$, and for every $\sigma \in \Sigma_2$, the map $y \mapsto Q(y, \sigma y)$ is Borel*

such that for a. e. $x \sim y \sim z \in X_2$:

$$c_1(\tau^{-1}x, \tau^{-1}y, \tau^{-1}z) = V(x, z)\alpha^2_{(z,y)}(Q(z, y))c_2(x, y, z)V(y, z)^*$$

Proof. First, let $\tilde{M}_1$ and $\tilde{M}_2$ be I-isomorphic and κ be the isomorphism as in the definition. In particular, $\kappa(I_1(A_1)) = I_2(A_2)$. Let $\tau : X_1 \to X_2$ be the corresponding Borel isomorphism (i. e. such that for $a \in A_1$: $\kappa(I_1(a)) = I_2(a\circ\tau^{-1})$). Such τ exists and satisfy $\mu_1 \sim \mu_2 \circ \tau$ by [10, Appendice IV]. Let

$$d\mu_2(\tau y) = r_{12}(\tau y)d\mu_1(y), \qquad y \in X_1$$

By Lemma 13.6, $\tau^{(2)}(R_1) = R_2$. Furthermore, the algebras M_1 and M_2 have cyclic and separable vectors, hence the isomorphism $I_2^{-1} \circ (\kappa|I_1(M_1)) \circ I_1 : M_1 \to M_2$ is spatial. Let it be implemented by a unitary operator $W : H_1 \to H_2$. Thus, for $a \in M_1$:

$$\kappa(I_1(a)) = I_2(WaW^{-1})$$

We have: for $a \in A_1$, $\kappa(I_1(a)) = I_2(a \circ \tau^{-1})$, hence for $a \in A_1$: $WaW^{-1} = a \circ \tau^{-1}$. By Lemma 13.7, there exist unitary operators $W(\tau y)$ from $H_1(y)$ onto $H_2(\tau y)$ such that for a. e. $x \in X_2$:

$$(Wf)(x) = r_{12}(x)^{-1/2} W(x) f(\tau^{-1}x) \qquad f \in H_1$$

Then for $a \in M_1$, $f \in H_2$, a. e. $x \in X_2$:

$$\begin{aligned}(WaW^*f)(x) &= r_{12}(x)^{-1/2} W(x)(aW^*f)(\tau^{-1}x)\\ &= r_{12}(x)^{-1/2} W(x) a(\tau^{-1}x)(W^*f)(\tau^{-1}x)\\ &= W(x)a(\tau^{-1}x)W(x)^* f(x)\end{aligned}$$

Thus, $(WaW^*)(x) = W(x)a(\tau^{-1}x)W(x)^*$, $a \in M_1$.

For $(x,y) \in R_2$, let $V_0(x,y) : H_1(\tau^{-1}y) \to H_2(y)$ be defined as follows:

$$V_0(x,y) = U^2_{(y,x)} W(x) U^1_{\tau^{-1}x,\tau^{-1}y)}$$

Let $V_0 : \tilde{H}_1 \to \tilde{H}_2$ be defined as follows:

$$(V_0 f)(x,y) = r_{12}(y)^{-1/2} V_0(x,y) f(\tau^{-1}x, \tau^{-1}y) \qquad f \in \tilde{H}_1$$

Then V_0 is unitary.

The straight calculation shows that

$$V_0 I_1(a) V_0^* = I_2(WaW^*) = \kappa(I_1(a))$$

Let $\tilde{M}_3 = V_0 \tilde{M}_1 V_0$. Then $I_2(M_2) = \kappa(I_1(M_1)) = V_0 I_1(M_1) V_0^*$ is the Cartan subalgebra of $\tilde{M}_3$.

For every $\sigma \in \Sigma_1$, $f \in \tilde{H}_2$ we have:

$$\begin{aligned}&(V_0 \pi_1(\sigma) V_0^* f)(x,y)\\ &= V_0(x,y) c_1(\tau^{-1}x, \sigma\tau^{-1}x, \tau^{-1}y) V_0(\tau\sigma\tau^{-1}x, y)^* f(\tau\sigma\tau^{-1}x, y)\end{aligned}$$

Let

$$c_3(x, \tau\sigma\tau^{-1}x, y) = V_0(x,y) c_1(\tau^{-1}x, \sigma\tau^{-1}x, \tau^{-1}y) V_0(\tau\sigma\tau^{-1}x, y)^*$$

Then c_3 is a "cocycle" and $\tilde{M}_3$ is the crossed product of M_2 by R_2 with this "cocycle".

Let λ be an isomorphism of $\tilde{M}_3$ onto $\tilde{M}_2$ defined by the formula

$$\lambda = \kappa \circ \operatorname{Ad}(V_0^*)$$

Then for $a \in M_2$:

$$\lambda(I_2(a)) = \kappa(V_0^* I_2(a) V_0) = \kappa(I_1(W^*aW)) = I_2(a)$$

Thus, $\tilde{M}_3$ and $\tilde{M}_2$ are $I_2(M_2)$-isomorphic. By Theorem 13.3, c_3 and c_2 are "cohomologous", i. e. for a. e. $x \sim y \sim z \in X_2$:

$$c_3(x,y,z) = V_1(x,z) \alpha^2_{(z,y)}(Q(x,y)) c_2(x,y,z) V_1(y,z)^*$$

where V_1 and Q are as in the beginning of Subsection 13.1. Hence,

$$\begin{aligned} &c_1(\tau^{-1}x,\tau^{-1}y,\tau^{-1}z)\\ &\quad= V_0(x,z)^*V_1(x,z)\alpha^2_{(z,y)}(Q(x,y))c_2(x,y,z)V_1(y,z)^*V_0(y,z) \end{aligned}$$

Let $V(x,z) = V_0(x,z)^*V_1(x,z)$. Then:

1. $c_1(\tau^{-1}x,\tau^{-1}y,\tau^{-1}z) = V(x,z)\alpha^2_{(z,y)}(Q(x,y))c^2(x,y,z)V(y,z)^*$;
2. $$\begin{aligned} &U^1_{(\tau^{-1}x,\tau^{-1}y)}V(x,y)U^2_{(y,x)} = U^1_{(\tau^{-1}x,\tau^{-1}y)}V_0(x,y)^*V_1(x,y)U^2_{(y,x)}\\ &= U^1_{(\tau^{-1}x,\tau^{-1}y)}U^1_{(\tau^{-1}y,\tau^{-1}x)}W(x)^*U^2_{(x,y)}V_1(x,y)U^2_{(y,x)}\\ &= W(x)^*U^2_{(x,y)}V_1(x,y)U^2_{(y,x)} = W(x)^*\alpha^2_{(x,y)}(V_1(x,y))\\ &= W(x)^*\bar{V}_1(x,y) \end{aligned}$$
   where $\bar{V}_1(x,y) = \alpha^2_{(x,y)}(V_1(x,y)) \in M_2'(y)$, so
   $\mathrm{Ad}\,(W(x)^*\bar{V}_1(x,y)) = \mathrm{Ad}\,(W(x)^*)$ — does not depend on y.

Conversely, assume that there exist τ, V and θ as in the formulation of the theorem.

Let $\kappa_x = \mathrm{Ad}\,(U^1_{(\tau^{-1}x,\tau^{-1}y)}V(x,y)U^2_{(y,x)})$.
By the conditions, κ_x is an isomorphism from $M_2(x)$ onto $M_1(\tau^{-1}x)$ not depending on y. But $M_2(x)$ and $M_1(\tau^{-1}x)$ possess cyclic and separating vectors. Hence this isomorphism is spatial. Let a unitary operator $W(x) : H_1(\tau^{-1}x) \to H_2(x)$ be such that $\kappa_x = \mathrm{Ad}\,(W(x)^*)$. Let $W : H_1 \to H_2$ be defined as follows:

$$(Wf)(x) = r_{12}(x)^{-1/2}W(x)f(\tau^{-1}x), \qquad f \in H_1, \quad x \in X_2$$

where r_{12} is as in the beginning of this proof. Then W is unitary, and for $a \in M_1$: $WaW^* \in M_2$.

For $(x,y) \in R_2$ let

$$\begin{aligned} V_0(x,y) &= U^2_{(y,x)}W(x)U^1_{(\tau^{-1}x,\tau^{-1}y)}\\ V_1(x,y) &= V(x,y)\cdot V_0(x,y) \end{aligned}$$

It follows from the properties of V and V_0 that $V_1(x,y) \in M_2'(y)$. Let

$$\begin{aligned} &c_3(x,y,z) = V_1(x,z)\alpha^2_{(z,y)}(Q(x,y))c_2(x,y,z)V_1(y,z)^*\\ &= V_0(x,z)c_1(\tau^{-1}x,\tau^{-1}y,\tau^{-1}z)V_0(y,z)^* \end{aligned}$$

Then c_3 is a "cocycle" which is "cohomologous" to c_2.

Let $\tilde{M}_3$ be the crossed product of M_2 by R_2 with the set of correspondence maps α^2 and the "cocycle" c_3. We will denote all the objects concerning $\tilde{M}_3$ by the subscript or superscript 3.

By Theorem 13.3, $\tilde{M}$ is $I_2(M_2)$-isomorphic to $\tilde{M}_2$. Let $\kappa_1 : \tilde{M}_3 \to \tilde{M}_2$ be the corresponding isomorphism.

Let V_0 be defined by:

$$(V_0f)(x,y) = r_{12}(y)^{-1/2}V_0(x,y)f(\tau^{-1}x,\tau^{-1}y) \qquad f \in \tilde{H}_1$$

Then $V_0 : \tilde{H}_1 \to \tilde{H}_2$ is unitary, and:

$$\begin{aligned} V_0 I_1(a) V_0^* &= I_2(WaW^*), \quad a \in M_1; \\ V_0 \pi_1(\sigma) V_0^* &= \pi_3(\tau\sigma\tau^{-1}), \quad \sigma \in \Sigma_1. \end{aligned}$$

So, $\operatorname{Ad} V_0$ is an isomorphism of $\tilde{M}_1$ onto $\tilde{M}_3$. Hence, $\kappa = \kappa_1 \circ \operatorname{Ad} V_0$ is an isomorphism of $\tilde{M}_1$ onto $\tilde{M}_2$. It is clear that κ carries $I_1(M_1)$ onto $I_2(M_2)$ and $I_1(A_1)$ onto $I_2(A_2)$.

□

The following lemma states some result which will be used in the following.

Lemma 13.10 *Let $\kappa : \tilde{M}_1 \to \tilde{M}_2$ be an I-isomorphism which is implemented by a unitary operator $W : \tilde{H}_1 \to \tilde{H}_2$ such that $W\tilde{J}_1 = \tilde{J}_2 W$.*

Let the isomorphism $\kappa | I_1(A_1) : I_1(A_1) \to I_2(A_2)$ correspond to a Borel isomorphism $\tau : X_1 \to X_2$.

Then for every $c \in C_1$, a. e. $(x, y) \in R_2$:

$$(WcW^*)(x, y) = c(\tau^{-1}x, \tau^{-1}y)$$

PROOF. It is enough to check the assertion for $c \in I_1(A_1)$ and for $c \in B_1 = \tilde{J}_1 I_1(A_1) \tilde{J}_1$. The checking is straightforward.

□

CHAPTER 14

BIMODULES AND SUBALGEBRAS OF $\tilde{M}$

NOTATION. Let for $(x,y) \in R$: $M(x,y) \equiv M(y)$.

Let

$$\tilde{M}_0 = \Big\{ \sum_{i=1}^{n} I(a_i)\pi(\tau_i) \mid n \in \mathbb{N}, a_i \in M, \tau_i \in \Sigma, i = 1, \ldots, n \Big\}$$

Let $\tilde{Z}$ be the following von Neumann algebra:

$$\tilde{Z} = \int_R Z(M(x,y))d\nu(x,y)$$

where $Z(M(x,y))$ is the center of $M(x,y)$.

For every projection $B \in \tilde{Z}$ let

$$\mathcal{T}(B) = \{T \in \tilde{M} \mid T(x,y)B[x,y] = T(x,y) \text{for a. e. } (x,y) \in R\}$$

where $T(x,y)$ are the coordinates of $T \in \tilde{M}$ as above, and $B[x,y]$ are the coordinates of $B \in \tilde{Z}$ in the decomposition

$$B = \int_R B[x,y]d\nu(x,y)$$

Let $\tilde{Z}^p$ be the set of all projections in $\tilde{Z}$.

For $B_1, B_2, B \in \tilde{Z}^p$ let

$$\begin{aligned} (B_1 \circ B_2)[x,y] &= \sup_{z \sim y}\Big(\alpha_{(y,z)}(B_1[x,z])\underbrace{c(x,z,y)B_2[z,y]c(x,z,y)^{-1}}_{\in Z(M(y))}\Big) \\ \theta(B)[x,y] &= c(x,y,y)\alpha_{(y,x)}(B[y,x])c(x,y,y)^{-1} \end{aligned}$$

Then $B_1 \circ B_2 \in \tilde{Z}^p$, $\theta(B) \in \tilde{Z}^p$.

REMARK. It is clear that $\mathcal{T}(B_1) \cap \mathcal{T}(B_2) = \mathcal{T}(\inf(B_1, B_2)) = \mathcal{T}(B_1 B_2)$.

Lemma 14.1 *For every $B \in \tilde{Z}^p$, $\mathcal{T}(B)$ is a σ-weakly closed subspace of $\tilde{M}$.*

PROOF. Let $T \in \tilde{M} \setminus \mathcal{T}(B)$. It is enough to prove that $T \notin \overline{\mathcal{T}(B)}^{\sigma w}$.

There exist $\tau \in \Sigma$ and $Q \subset X$ with $\mu(Q) > 0$ such that for $x \in Q$: $B[\tau x, x]T(\tau x, x) \neq T(\tau x, x)$. Let

$$t(x) = T(\tau x, x), \qquad t = \int_X t(x) d\mu(x);$$

$$b(x) = B[\tau x, x], \qquad b = \int_X b(x) d\mu(x)$$

Then $t, b \in M$, b is a central projection. By the assumption, $bt \neq t$. Hence, $bt\phi_0 \neq t\phi_0$. Thus, $t\phi_0 \notin BH = ((BH)^\perp)^\perp$. So, there exists $g \in H$ such that $bg = 0$, but $\langle t\phi_0, g\rangle \neq 0$. Let $\tilde{f}, \tilde{g} \in \tilde{H}$ be defined as follows:

$$\tilde{g}(x,y) = \begin{cases} g(y) & \text{if } x = \tau y \\ 0 & \text{otherwise} \end{cases}, \qquad \tilde{f}(x,y) = \begin{cases} c(\tau y, y, y)^{-1}\phi_0(y) & \text{if } x = y \\ 0 & \text{otherwise} \end{cases}$$

It is easy to see that $\langle T\tilde{f}, \tilde{g}\rangle \neq 0$, but for every $S \in \mathcal{T}(B)$: $\langle S\tilde{f}, \tilde{g}\rangle = 0$.

Thus, the functional $\langle (\cdot)\tilde{f}, \tilde{g}\rangle$ separates T from $\mathcal{T}(B)$. Hence, $T \notin \overline{\mathcal{T}(B)}^{\sigma w}$. □

Lemma 14.2 *For $B \in \tilde{Z}^p$, $\mathcal{T}(B)$ is a bimodule with respect to $I(M)$.*

PROOF. The proof is straightforward. □

Theorem 14.3 *Let $B \in \tilde{Z}^p$.*

1. *$\mathcal{T}(B)$ is a subalgebra if and only if $B \circ B \leq B$;*
2. *$\mathcal{T}(B)$ is self-adjoint if and only if $B = \theta(B)$;*
3. *$1_{\tilde{H}} \in \mathcal{T}(B)$ if and only if $B \geq P_\triangle$, where $P_\triangle \in \tilde{Z}^p$, $P_\triangle[x,y] = \delta_{(x,y)} 1_{H(y)}$;*
4. *$\mathcal{T}(B) \cap \mathcal{T}(B)^* = I(M)$ if and only if $B \cdot \theta(B) = P_\triangle$;*
5. *$\tilde{M}$ is generated (as a σ-weakly closed bimodule) by $\mathcal{T}(B) + \mathcal{T}(B)^*$ if and only if $\sup(B, \theta(B)) = 1_{\tilde{H}}$.*

PROOF.

1. It is enough to prove that $(B_1 \circ B_2)(x,y) \leq B(x,y)$ for a. e. $(x,y) \in R$ if and only if $\mathcal{T}(B_1)\mathcal{T}(B_2) \subseteq \mathcal{T}(B)$ (here $B_1, B_2, B \in \tilde{Z}^p$). It be checked immediately.
2. It is enough to prove that $\mathcal{T}(B)^* = \mathcal{T}(\theta(B))$.

$$T \in \mathcal{T}(B) \iff T(x,y)B[x,y] = T(x,y) \text{ a. e.}$$

$$\iff T(x,y)^* B[x,y] = T(x,y)^* \text{ a. e.}$$

$$\iff T^*(x,y)\theta(B)[x,y]$$

$$= c(x,y,y)\alpha_{(y,x)}(T(y,x)^*)c(x,y,y)^{-1}\cdot$$
$$c(x,y,y)\alpha_{(y,x)}(B[y,x])c(x,y,y)^{-1}$$
$$= c(x,y,y)\alpha_{(y,x)}(T(y,x)^*B[y,x])c(x,y,y)^{-1}$$
$$= c(x,y,y)\alpha_{(y,x)}(T(y,x)^*)c(x,y,y)^{-1} = T^*(x,y) \text{ a. e.}$$
$$\Longleftrightarrow T^* \in \mathcal{T}(\theta(B))$$

3. It follows from the definitions.

4. We have only to check that $I(M) = \mathcal{T}(P_\triangle)$.
For every $a \in M$, $I(a) \in \mathcal{T}(P_\triangle)$ by the definitions.
Conversely, let $T \in \mathcal{T}(P_\triangle)$. Then T acts as follows:

$$(T\xi)(x,y) = \alpha_{(y,x)}(T(x,x))\xi(x,y), \qquad \xi \in \tilde{H}$$

Take $a(x) = T(x,x)$, $a = \int_X a(x)d\mu(x)$. Then $a \in M$ and $T = I(a) \in I(M)$.

5. By the proof of part 2, $\mathcal{T}(B)^* = \mathcal{T}(\theta(B))$. Thus, it is enough to prove that for $B_1, B_2 \in \tilde{Z}^p$, the algebra $\tilde{M}$ is generated by $\mathcal{T}(B_1) + \mathcal{T}(B_2)$ if and only if $\sup(B_1, B_2) = 1_{\tilde{H}}$.

   (a) Assume first that $\sup(B_1, B_2) \neq 1_{\tilde{H}}$. Write $B_3 = \sup(B_1, B_2)$, $B_4 = 1_{\tilde{H}} - B_3$. Then $B_4 \neq 0$. There exists $\tau_0 \in \Sigma$ such that

$$\mu(\{x \in X \mid B_4[x, \tau_0 x] \neq 0\}) > 0$$

   Let $a_0(x) = \alpha_{(x,\tau_0 x)}(B_4[x, \tau_0 x])$, $T_0 = I(a_0)\pi(\tau_0)$. Then

$$T_0(x,y) = \begin{cases} B_4[x,\tau_0 x] & \text{if } y = \tau_0 x \\ 0 & \text{otherwise} \end{cases}$$

   Hence, $T_0 \in \mathcal{T}(B_4)$. Therefore,

$$T_0(x,y)B_3[x,y] = T_0(x,y)B_4[x,y]B_3[x,y]$$
$$= T_0(x,y)(1_{H(y)} - B_3[x,y])B_3[x,y] = 0 \neq T_0(x,y)$$

   for every (x,y) in some set of positive ν-measure. Thus, $T_0 \notin \mathcal{T}(B_3)$. On the other hand, $B_1 \leq B_3$ and $B_2 \leq B_3$, hence $\mathcal{T}(B_1) \subseteq \mathcal{T}(B_3)$ and $\mathcal{T}(B_2) \subseteq \mathcal{T}(B_3)$, so the σ-weakly closed bimodule generated by $\mathcal{T}(B_1) + \mathcal{T}(B_2)$ is contained in $\mathcal{T}(B_3)$. Therefore it does not contain T_0 hence it is not equal to $\tilde{M}$.

   (b) Now assume that $\sup(B_1, B_2) = 1_{\tilde{H}}$. We show that $\mathcal{T}(B_1) + \mathcal{T}(B_2)$ generated $\tilde{M}$ as a σ-weakly closed subspace. The set $\tilde{M}_0$ (see the notation in the beginning of this section) is clearly an algebra which is dense in $\tilde{M}$. It is enough to prove that $\tilde{M}_0 \subseteq \mathcal{T}(B_1) + \mathcal{T}(B_2)$. (It follows from this that $\tilde{M} \subseteq \overline{\mathcal{T}(B_1) + \mathcal{T}(B_2)}^{\sigma w}$.)
   Let $T \in \tilde{M}_0$. Assume first that T is of the form

$$T = I(a)\pi(\tau), \qquad \text{with} \quad a \in M, \quad \tau \in \Sigma$$

Then

$$T(x,y) = \begin{cases} \alpha_{(\tau x,x)}(a(x)) & \text{if } y = \tau x \\ 0 & \text{otherwise} \end{cases}$$

Let $a_1(x) = a(x) \cdot \alpha_{(x,\tau x)}(B_1[x.,\tau x])$ and $a_1 = \int_X a_1(x) d\mu(x)$. Then $a_1 \in M$; write $a_2 = a - a_1$.
Let $T_1 = I(a_1)\pi(\tau)$, $T_2 = I(a_2)\pi(\tau)$. Then $T_1, T_2 \in \tilde{M}$ and

$$T_1 + T_2 = I(a_1 + a_2)\pi(\tau) = I(a)\pi(\tau) = T$$

We have:

$$T_1(x,y) = \begin{cases} \alpha_{(\tau x,x)}(a_1(x)) & \text{if } y = \tau x \\ 0 & \text{otherwise} \end{cases}$$

$$= \begin{cases} \alpha_{(\tau x,x)}(a(x)B_1[x,\tau x]) & \text{if } y = \tau x \\ 0 & \text{otherwise} \end{cases}$$

$$= T(x,y)B_1[x,y] = T(x,y)B_1[x,y]B_1[x,y] = T_1(x,y)B_1[x,y]$$

Thus, $T_1 \in \mathcal{T}(B_1)$.
Futhermore, it is easy to see that $T_2 \in \mathcal{T}(B_2)$. So $T = T_1 + T_2 \in \mathcal{T}(B_1) + \mathcal{T}(B_2)$.
Now, let $T \in \tilde{M}_0$ be arbitrary, i. e. $T = \sum_{i=1}^n T_i$, $T_i = I(a_i)\pi(\tau_i)$, $a_i \in M$, $\tau_i \in \Sigma$, $i = 1, \ldots, n$. As was shown above, $T_i \in \mathcal{T}(B_1) + \mathcal{T}(B_2)$ for every $i = 1, \ldots, n$, hence $T \in \mathcal{T}(B_1) + \mathcal{T}(B_2)$.

□

Theorem 14.4 (See [25, Theorem 3.4].) *Let $B \in \tilde{Z}^p$ be such that $B \circ B = B$, $\theta(B) = B$ and $B[x,x] = 1_{H(x)}$ for a. e. $x \in X$. Then $\mathcal{T}(B)$ is a von Neumann subalgebra of $\tilde{M}$, and there exists a unique conditional expectation Φ from $\tilde{M}$ onto $\mathcal{T}(B)$ which preserves the state $\langle (\cdot)\tilde{\phi}_0, \tilde{\phi}_0 \rangle$, and this expectation is given by the formula*

$$\Phi(T)(x,y) = T(x,y)B[x,y], \qquad T \in \tilde{M}, \quad (x,y) \in R$$

PROOF. It was proved that $\mathcal{T}(B)$ is a subalgebra of $\tilde{M}$. By the conditions, $\mathcal{T}(B)$ contains $I(M)$.

1. The modular operator $\tilde{\Delta}$, associated with $\tilde{M}$ and $\tilde{\phi}_0$, commutes with $I(Z(M)) \cup I'(Z(M))$ by Proposition 9.23(1). In particular, for $a \in Z(M)$, $\xi \in \tilde{H}$, a. e. $(x,y) \in R$:

$$\alpha_{(y,x)}(a(x))\tilde{\Delta}(x,y)\xi(x,y) = \alpha_{(y,x)}(a(x))(\tilde{\Delta}\xi)(x,y)$$
$$= (I(a)\tilde{\Delta}\xi)(x,y) = (\tilde{\Delta}I(a)\xi)(x,y)$$
$$= \tilde{\Delta}(x,y)(I(a)\xi)(x,y) = \tilde{\Delta}(x,y)\alpha_{(y,x)}(a(x))\xi(x,y)$$

This is true for every $\xi \in \tilde{H}$, hence for a. e. $(x,y) \in R$:

$$\alpha_{(y,x)}(a(x))\tilde{\Delta}(x,y) = \tilde{\Delta}(x,y)\alpha_{(y,x)}(a(x))$$

This, in turn, is true for every $a \in Z(M)$. Hence $\tilde{\Delta}(x,y)$ commutes with $Z(M(y))$. So for $t \in \mathbb{R}$: $\tilde{\Delta}^{it}(x,y) = \tilde{\Delta}(x,y)^{it}$ commutes with $Z(M(y))$ too.

2. Let $T \in \mathcal{T}(B)$, $\xi \in \tilde{H}$, $(x, y) \in R$, $t \in \mathbb{R}$, $S = \sigma_t(T)$. (Here $\{\sigma_t\}$ is the modular group.) Then:

$$\begin{aligned}
&(S\xi)(x,y) = (\tilde{\Delta}^{it} T \tilde{\Delta}^{-it}\xi)(x,y) = \tilde{\Delta}^{it}(x,y)(T\tilde{\Delta}^{-it}\xi)(x,y) \\
&= \tilde{\Delta}^{it}(x,y) \textstyle\sum_z \alpha_{(y,z)}(T(x,z))c(x,z,y)(\tilde{\Delta}^{-it}\xi)(z,y) \\
&= \tilde{\Delta}^{it}(x,y) \textstyle\sum_z \alpha_{(y,z)}(T(x,z))c(x,z,y)\tilde{\Delta}^{-it}(z,y)\xi(z,y) \\
&= \textstyle\sum_z \tilde{\Delta}^{it}(x,y)\alpha_{(y,z)}(T(x,z))c(x,z,y)\tilde{\Delta}^{-it}(z,y)c(x,z,y)^{-1} \\
&\qquad\qquad\qquad\qquad\qquad\qquad\qquad\qquad\qquad c(x,z,y)\xi(z,y)
\end{aligned}$$

On the other hand,

$$(S\xi)(x,y) = \sum_z \alpha_{(y,z)}(S(x,z))c(x,z,y)\xi(z,y)$$

This is true for every $\xi \in \tilde{H}$. Hence,

$$\begin{aligned}
&\tilde{\Delta}^{it}(x,y)\alpha_{(y,z)}(T(x,z))c(x,z,y)\tilde{\Delta}^{-it}(z,y)c(x,z,y)^{-1} \\
&= \alpha_{(y,z)}(S(x,z)); \\
&S(x,z) = \alpha_{(z,y)}(\alpha_{(y,z)}(S(x,z))) \\
&= \alpha_{(z,y)}(\tilde{\Delta}^{it}(x,y)\alpha_{(y,z)}(T(x,z))c(x,z,y)\tilde{\Delta}^{-it}(z,y)c(x,z,y)^{-1}) \\
&= \alpha_{(z,y)}(\tilde{\Delta}^{it}(x,y))T(x,z)\alpha_{(z,y)}(c(x,z,y)\tilde{\Delta}^{-it}(z,y)c(x,z,y)^{-1})
\end{aligned}$$

Thus, the last expression does not depend on y. Let $y = z$ and get:

$$S(x,z) = \tilde{\Delta}^{it}(x,z)T(x,z)c(x,z,z)\tilde{\Delta}^{-it}(z,z)c(x,z,z)^{-1}$$

Hence:

$$\begin{aligned}
&B[x,z]S(x,z) \\
&= B[x,z]\tilde{\Delta}^{it}(x,z)T(x,z)c(x,z,z)\tilde{\Delta}^{-it}(z,z)c(x,z,z)^{-1} \\
&= \tilde{\Delta}^{it}(x,z)B[x,z]T(x,z)c(x,z,z)\tilde{\Delta}^{-it}(z,z)c(x,z,z)^{-1} \\
&= \tilde{\Delta}^{it}(x,z)T(x,z)c(x,z,z)\tilde{\Delta}^{-it}(z,z)c(x,z,z)^{-1} = S(x,z)
\end{aligned}$$

So, $S \in \mathcal{T}(B)$. Thus, the modular automorphism group $\{\sigma_t\}$ leaves $\mathcal{T}(B)$ invariant.

3. By a theorem of Takesaki [34, Theorem on page 309], there exists a faithful normal conditional expectation Φ from $\tilde{M}$ onto $\mathcal{T}(B)$, which preserves the state $\langle(\cdot)\tilde{\phi}_0, \tilde{\phi}_0\rangle$.

   Let Φ_1, Φ_2 be two such conditional expectations. Then for $T \in \tilde{M}$, $a \in \mathcal{T}(B)$:

$$\begin{aligned}
&\langle \Phi_1(T)\tilde{\phi}_0, a\tilde{\phi}_0\rangle = \langle a^*\Phi_1(T)\tilde{\phi}_0, \tilde{\phi}_0\rangle = \langle \Phi_1(a^*T)\tilde{\phi}_0, \tilde{\phi}_0\rangle \\
&= \langle a^*T\tilde{\phi}_0, \tilde{\phi}_0\rangle = \langle \Phi_2(a^*T)\tilde{\phi}_0, \tilde{\phi}_0\rangle = \langle a^*\Phi_2(T)\tilde{\phi}_0, \tilde{\phi}_0\rangle \\
&\qquad\qquad\qquad\qquad\qquad\qquad\qquad\qquad = \langle \Phi_2(T)\tilde{\phi}_0, a\tilde{\phi}_0\rangle
\end{aligned}$$

This is true for every $a \in \mathcal{T}(B)$. Thus, $\Phi_1(T)\tilde{\phi}_0 - \Phi_2(T)\tilde{\phi}_0$ is orthogonal to $K = \overline{\{a\tilde{\phi}_0 \mid a \in \mathcal{T}(B)\}}^{\|\cdot\|}$. But, on the other hand, $\Phi_1(T)\tilde{\phi}_0 - \Phi_2(T)\tilde{\phi}_0 \in K$, hence $\Phi_1(T)\tilde{\phi}_0 - \Phi_2(T)\tilde{\phi}_0 = 0$ and $\Phi_1(T)\tilde{\phi}_0 = \Phi_2(T)\tilde{\phi}_0$. Since $\tilde{\phi}_0$ is separating for $\tilde{M}$, $\Phi_1(T) = \Phi_1(T)$.

4. It remains to check that for $T \in \tilde{M}$:

$$\Phi(T)(x,y) = T(x,y)B[x,y]$$

(a) First, let $T = \pi(\tau)$ with $\tau \in \Sigma$. For $a \in M$:

$$\begin{aligned}
&I(a)T^*\Phi(T) = I(a)\pi(\tau)^*\Phi(\pi(\tau)) \\
&= \pi(\tau)^*\underbrace{\pi(\tau)I(a)\pi(\tau)^*}_{\in I(M)\subset \mathcal{T}(B)}\Phi(\pi(\tau)) \\
&= \pi(\tau)^*\Phi(\pi(\tau)I(a)\pi(\tau)^*\pi(\tau)) \\
&= \pi(\tau)^*\Phi(\pi(\tau)I(a)) = \pi(\tau)^*\Phi(\pi(\tau))I(a) = T^*\Phi(T)I(a)
\end{aligned}$$

Thus, $T^*\Phi(T) \in I(M)'$. In particular, $T^*\Phi(T) \in I(A)' \cap \tilde{M} = I(M)$. Hence, $T^*\Phi(T) \in I(M)' \cap I(M) = I(Z(M))$. So $\Phi(\pi(\tau)) = \pi(\tau)I(b_\tau)$ for some $b_\tau \in Z(M)$. In particular,

$$\begin{aligned}
&\pi(\tau)I(b_\tau) = \Phi(\pi(\tau)) = \Phi(\Phi(\pi(\tau))) \\
&= \Phi(\pi(\tau)I(b_\tau)) = \Phi(\pi(\tau))I(b_\tau) = \pi(\tau)I(b_\tau)I(b_\tau) = \pi(\tau)I(b_\tau^2)
\end{aligned}$$

Thus, $b_\tau = b_\tau^2$. Furthermore, $b_\tau \in Z(M)$, hence b_τ commutes with b_τ^*, hence b_τ is a projection. We have:

$$\begin{aligned}
&\Phi(\pi(\tau))(x,y) = (\pi(\tau)I(b_\tau))(x,y) \\
&= \begin{cases} c(x,\tau x,\tau x)b_\tau(\tau x)c(x,\tau x,\tau x)^{-1} & \text{if } y = \tau x \\ 0 & \text{otherwise} \end{cases}
\end{aligned}$$

But $\Phi(\pi(\tau)) \in \mathcal{T}(B)$, hence

$$c(x,\tau x,\tau x)b_\tau(\tau x)c(x,\tau x,\tau x)^{-1} \leq B[x,\tau x] \tag{14.1}$$

Now, let $d_\tau(\tau x) = c(x,\tau x,\tau x)^{-1}B[x,\tau x]c(x,\tau x,\tau x)$ for $x \in X$, and $d_\tau = \int_X d_\tau(y)d\mu(y)$. Then d_τ is a projection in $Z(M)$, and

$$\pi(\tau)I(d_\tau)(x,y) = \begin{cases} B[x,\tau x] & \text{if } y = \tau x \\ 0 & \text{otherwise} \end{cases}$$

Hence $\pi(\tau)I(d_\tau) \in \mathcal{T}(B)$ and

$$\pi(\tau)I(b_\tau)I(d_\tau) = \Phi(\pi(\tau))I(d_\tau) = \Phi(\pi(\tau)I(d_\tau)) = \pi(\tau)I(d_\tau)$$

Thus, $b_\tau d_\tau = d_\tau$, i. e. $d_\tau \leq b_\tau$. Hence for a. e. $x \in X$: $d_\tau(\tau x) \leq b_\tau(\tau x)$, and

$$c(x,\tau x,\tau x)d_\tau(\tau x)c(x,\tau x,\tau x)^{-1} \leq c(x,\tau x,\tau x)b_\tau(\tau x)c(x,\tau x,\tau x)^{-1}$$

In other words:

$$B[x,\tau x] \le c(x,\tau x,\tau x)b_\tau(\tau x)c(x,\tau x,\tau x)^{-1} \tag{14.2}$$

It follows from (14.1) and (14.2) that

$$c(x,\tau x,\tau x)b_\tau(\tau x)c(x,\tau x,\tau x)^{-1} = B[x,\tau x]$$

and

$$\Phi(\pi(\tau))(x,y) = \begin{cases} B[x,\tau x] & \text{if } y=\tau x \\ 0 & \text{otherwise} \end{cases} = \pi(\tau)(x,y)B[x,y]$$

(b) Now, let $T \in \tilde{M}_0$, $T = \sum_{i=1}^n I(a_i)\pi(\tau_i)$ $a_i \in M$, $\tau_i \in \Sigma$, $i = 1,\dots,n$. The result can be got from the part (4a) of this proof by summation.

(c) Finally, let $T \in \tilde{M}$ be arbitrary. There exists a net $(T_\alpha) \subset \tilde{M}_0$ such that $T_\alpha \overset{\sigma s}{\to} T$. In particular, $T_\alpha\tilde{\phi}_0 \to T\tilde{\phi}_0$. There exists some subsequence (T_n) of (T_α) such that $T_n\tilde{\phi}_0 \to T\tilde{\phi}_0$. By [10, Ch. II, §1, n. 5, Proposition 5(ii)], for some subsequence (T_{n_m}) of (T_n):

$$(T_{n_m}\tilde{\phi}_0)(x,y) \to (T\tilde{\phi}_0)(x,y) \qquad \text{a. e.}$$

Thus, for a. e. $(x,y) \in R$:

$$T_{n_m}(x,y)c(x,y,y)\tilde{\phi}_0(y,y) \to T(x,y)c(x,y,y)\tilde{\phi}_0(y,y)$$

Hence, for a. e. $(x,y) \in R$:

$$\begin{aligned}
&(\Phi(T_{n_m})\tilde{\phi}_0)(x,y) = \Phi(T_{n_m})(x,y)c(x,y,y)\tilde{\phi}_0(x,y)\\
&= T_{n_m}(x,y)B[x,y]c(x,y,y)\tilde{\phi}_0(x,y)\\
&= B[x,y]T_{n_m}(x,y)c(x,y,y)\tilde{\phi}_0(x,y)\\
&\to B[x,y]T(x,y)c(x,y,y)\tilde{\phi}_0(x,y) = T(x,y)B[x,y]c(x,y,y)\tilde{\phi}_0(x,y)
\end{aligned}$$

On the other hand, by [32, Subsection 9.2] Φ is σ-strongly continous hence $\Phi(T_\alpha) \overset{\sigma s}{\to} \Phi(T)$,hence $\Phi(T_\alpha)\tilde{\phi}_0 \to \Phi(T)\tilde{\phi}_0$, hence $\Phi(T_{n_m})\tilde{\phi}_0 \to \Phi(T)\tilde{\phi}_0$.

By [10, Ch. II, §1, n. 5, Proposition 5(ii)], for some subsequence $(T_{n_{m_k}})$ of (T_{n_m}):

$$(\Phi(T_{n_{m_k}})\tilde{\phi}_0)(x,y) \to (\Phi(T)\tilde{\phi}_0)(x,y)$$

Thus, for a. e. $(x,y) \in R$:

$$(\Phi(T)\tilde{\phi}_0)(x,y) = T(x,y)B[x,y]c(x,y,y)\tilde{\phi}_0(x,y)$$

Hence, for a. e. $(x,y) \in R$:

$$\begin{aligned}
&\Phi(T)(x,y)c(x,y,y)\tilde{\phi}_0(y,y) = (\Phi(T)\tilde{\phi}_0)(x,y)\\
&= T(x,y)B[x,y]c(x,y,y)\tilde{\phi}_0(y,y);
\end{aligned}$$

$$
\begin{aligned}
&c(x,y,y)^{-1}\Phi(T)(x,y)c(x,y,y)\tilde{\phi}_0(y,y)\\
&= c(x,y,y)^{-1}T(x,y)B[x,y]c(x,y,y)\tilde{\phi}_0(y,y)
\end{aligned}
$$

But $\tilde{\phi}_0(y,y) = \phi_0(y)$ is a cyclic and separating vector for a. e. $y \in X$. Hence for a. e. $(x,y) \in R$:

$$\Phi(T)(x,y) = T(x,y)B[x,y]$$

□

CHAPTER 15

SPECTRAL THEOREM FOR BIMODULES

Let $\tilde{M}$ be the crossed product of $M = \int_X M(x)d\mu(x)$ by R.

In this section we shall see that $I(M)$-bimodules in $\tilde{M}$ have a special form which generalizes Theorem 2.5 in [25]. This will apply, in particular, to subalgebras of $\tilde{M}$ that contain $I(M)$.

In what follows, "a bimodule in $\tilde{M}$" means a bimodule with respect to $I(M)$.

For the proof of the main result we will have to assume that R is hyperfinite (see [14]).

Definition 15.1 (See [14, Section 4].) *An equivalence relation R on a set X is called* **hyperfinite**, *if there are equivalence relations R_n on X with finite cosets such that $R_n \subseteq R_{n+1}$ for $n = 1, 2, \dots$ and $R = \bigcup_{n=1}^{\infty} R_n$.*

NOTATION. For every $I(M)$-bimodule $\mathcal{S}$ define $\mathcal{P}_\mathcal{S} \subset \tilde{Z}^p$ and $P_\mathcal{S} \in \tilde{Z}^p$ as follows:

$$\begin{aligned} \mathcal{P}_\mathcal{S} &= \{P \in \tilde{Z}^p \mid \mathcal{S} \subseteq \mathcal{T}(P)\} \\ P_\mathcal{S} &= \inf \mathcal{P}_\mathcal{S} \end{aligned}$$

From now on, let $\mathcal{S}$ be a σ-weakly closed $I(M)$-bimodule in $\tilde{M}$.

Lemma 15.2 *$\mathcal{P}_\mathcal{S}$ is a decreasing net with respect to the natural order, where the index of every element equals to this element itself, and the order of indices is opposite to the natural order.*

PROOF. One has only to check that $\mathcal{P}_\mathcal{S}$ is a net with this order of indices.

Let $P_1, P_2 \in \mathcal{P}_\mathcal{S}$, $T \in \mathcal{S}$. Then $T \in \mathcal{T}(P_1) \cap \mathcal{T}(P_2)$, hence for a. e. $(x, y) \in R$:

$$\begin{aligned} &T(x,y)(P_1P_2)[x,y] = T(x,y)P_1[x,y]P_2[x,y] \\ &= (T(x,y)P_1[x,y])P_2[x,y] = T(x,y)P_2[x,y] = T(x,y) \end{aligned}$$

Thus, $T \in \mathcal{T}(P_1P_2)$. This holds for every $T \in \mathcal{S}$, hence $\mathcal{S} \subseteq \mathcal{T}(P_1P_2)$ and $P_1P_2 \in \mathcal{P}_\mathcal{S}$. By the order of indices above, P_1P_2 is preceded by P_1 and P_2. Thus, $\mathcal{P}_\mathcal{S}$ is a net.

□

Corollary 15.3 *$\mathcal{P}_\mathcal{S}$ strongly converges to $P_\mathcal{S}$.*

PROOF. Let $\mathcal{P}'_\mathcal{S} = \{1_{\tilde{H}} - P \mid P \in \mathcal{P}_\mathcal{S}\}$, $P'_\mathcal{S} = 1_{\tilde{H}} - P_\mathcal{S}$. Then $P'_\mathcal{S} = \sup \mathcal{P}'_\mathcal{S}$.

Let every $Q \in \mathcal{P}'_\mathcal{S}$ be equipped with an index with equals to $(1_{\tilde{H}} - Q) \in \mathcal{P}_\mathcal{S}$, with the order of indices as in Lemma 15.2. By Lemma 15.2, $\mathcal{P}'_\mathcal{S}$ is an increasing net. By [33, Subsection 2.16], $\mathcal{P}'_\mathcal{S}$ strongly converges to $P'_\mathcal{S}$, hence $\mathcal{P}_\mathcal{S} = 1_{\tilde{H}} - \mathcal{P}'_\mathcal{S}$ strongly converges to $P_\mathcal{S} = 1_{\tilde{H}} - P'_\mathcal{S}$.

□

Lemma 15.4 *There exists a subsequence* $(P_n)_{n=1}^{\infty}$ *of the net* $\mathcal{P}_S$ *such that* $P_n \to P_S$ *strongly.*

PROOF. By [10, Ch. I, §3, n. 1, Proposition 1] the strong topology on unit ball is metrisable.

□

Corollary 15.5 *There exists a subsequence* $(P_k)_{k=1}^{\infty} \subset \mathcal{P}_S$ *such that for a. e.* $(x,y) \in R$*:* $P_k[x,y] \to P[x,y]$ *strongly.*

PROOF. Let $(P_n)_{n=1}^{\infty}$ be as in Lemma 15.4. Then $P_n \to P_S$ strongly. By [10, Ch. II, §2, n. 3, Proposition 4], there exists a subsequence $(P_k)_{k=1}^{\infty}$ of $(P_n)_{n=1}^{\infty}$ such that $P_k[x,y] \to P_S[x,y]$ strongly a. e.

□

Theorem 15.6 *In the notations above:*

$$\mathcal{S} \subseteq \mathcal{T}(P_S)$$

PROOF. Let $(P_k)_{k=1}^{\infty} \subset \mathcal{P}_S$ be such that $P_k[x,y] \to P[x,y]$ strongly for $(x,y) \notin N_0$, $\nu(N_0) = 0$. Let $T \in \mathcal{S}$. For every $k \in \mathbb{N}$: $P_k \in \mathcal{P}_S$, hence

$$T(x,y)P_k[x,y] = T(x,y)$$

for $(x,y) \notin N_k$, $\nu(N_k) = 0$. Let $N = \bigcup_{k=0}^{\infty} N_k$. Then $\nu(N) = 0$ and for $(x,y) \notin N$:

$$T(x,y)P_k[x,y] = T(x,y), \qquad k = 1, 2, \ldots$$

In the limit we get

$$T(x,y)P_S[x,y] = T(x,y) \qquad (x,y) \notin N$$

Thus, $T \in \mathcal{T}(P_S)$. This is true for every $T \in \mathcal{S}$, hence $\mathcal{S} \subseteq \mathcal{T}(P_S)$.

□

REMARK. Theorem 15.6 holds for the general case, not only for the hyperfinite case!

NOTATION. For every $\psi \in H$, write as above:

$$\tilde{\psi}(x,y) = \begin{cases} 0 & \text{if } x \neq y \\ \psi(x) & \text{if } x = y \end{cases}$$

Then $\tilde{\psi} \in \tilde{H}$ and $\|\tilde{\psi}\|_{\tilde{H}} = \|\psi\|_H$.

Lemma 15.7 *For* $T \in \tilde{M}$, $f, g \in H$*:*

$$\langle T\tilde{f}, \tilde{g}\rangle = \langle E(T)f, g\rangle_H$$

PROOF. The proof is straightforward.

□

Lemma 15.8 *For every $T \in \tilde{M}$:*

$$I(E(T)) \in \overline{\text{conv}}^{\sigma w}\{UTU^* \mid U \in I(A), U \text{ is unitary}\}$$

PROOF. Let $\mathcal{U}$ be the group of all unitary operators in $I(A)$ with the σ-weak topology (which coincides with the weak and the strong topologies). This group is compact; let dU be the Haar measure on this group with total mass 1. For $T \in \tilde{M}$ write:

$$S_T = \int_{\mathcal{U}} UTU^* dU$$

By [5, Ch. VI, §1, n. 3, Proposition 9 et Corollaire], $S_T \in B(\tilde{H})$. Now, let $T \in \tilde{M}$, $T' \in \tilde{M}'$, $f, g \in \tilde{H}$.

$$\begin{aligned}
&\langle S_T T' f, g\rangle = \omega_{T'f,g}\left(\textstyle\int_{\mathcal{U}} UTU^* dU\right)\\
&= \textstyle\int_{\mathcal{U}} \omega_{T'f,g}(UTU^*)dU = \int_{\mathcal{U}} \langle UTU^*T'f, g\rangle dU\\
&= \textstyle\int_{\mathcal{U}} \langle T'UTU^* f, g\rangle dU = \int_{\mathcal{U}} \langle UTU^* f, T'^* g\rangle dU\\
&= \textstyle\int_{\mathcal{U}} \omega_{f,T'^*g}(UTU^*)dU = \omega_{f,T'^*g}\left(\int_{\mathcal{U}} UTU^* dU\right)\\
&= \langle S_T f, T'^* g\rangle = \langle T' S_T f, g\rangle
\end{aligned}$$

Thus, $S_T T' = T' S_T$ for every $T' \in \tilde{M}'$, hence $S_T \in \tilde{M}$.

Futhermore, let $V \in \mathcal{U}$, $f, g \in \tilde{H}$.

$$\begin{aligned}
&\langle S_T V f, g\rangle = \omega_{Vf,g}\left(\textstyle\int_{\mathcal{U}} UTU^* dU\right)\\
&= \textstyle\int_{\mathcal{U}} \omega_{Vf,g}(UTU^*)dU = \int_{\mathcal{U}} \langle UTU^* V f, g\rangle dU\\
&= \textstyle\int_{\mathcal{U}} \omega_{f,g}(UTU^* V)dU = \omega_{f,g}\left(\int_{\mathcal{U}} UTU^* V dU\right)\\
&= \left\langle \left(\textstyle\int_{\mathcal{U}} UTU^* V dU\right) f, g\right\rangle
\end{aligned}$$

Thus, $S_T V = \int_{\mathcal{U}} UTU^* V dU$ and, by a similar calculation,
$V S_T = \int_{\mathcal{U}} VUTU^* dU$ for $V \in \mathcal{U}$.

Let $U \in \mathcal{U}$ be fixed, $W = VU$. Then $dU = dW$, $U = V^* W$, $U^* = W^* V$, and

$$\begin{aligned}
&V S_T = \textstyle\int_{\mathcal{U}} VUTU^* dU\\
&= \textstyle\int_{\mathcal{U}} WTW^* V dW = \int_{\mathcal{U}} UTU^* V dU = S_T V
\end{aligned}$$

Thus, S_T commutes with $\mathcal{U}$ and, thus, with $I(A)$. Hence $S_T \in \tilde{M} \cap I(A)' = I(M)$, so $I(E(S_T)) = S_T$.

On the other hand, for $\psi \in H$:

$$\begin{aligned}
&\langle E(S_T)\phi_0, \psi\rangle_H = \langle S_T \tilde{\phi}_0, \tilde{\psi}\rangle\\
&= \omega_{\tilde{\phi}_0,\tilde{\psi}}\left(\textstyle\int_{\mathcal{U}} UTU^* dU\right) = \int_{\mathcal{U}} \omega_{\tilde{\phi}_0,\tilde{\psi}}(UTU^*)dU\\
&= \textstyle\int_{\mathcal{U}} \langle UTU^* \tilde{\phi}_0, \tilde{\psi}\rangle dU = \int_{\mathcal{U}} \langle E(UTU^*)\phi_0, \psi\rangle_H dU\\
&= \textstyle\int_{\mathcal{U}} \langle UE(T)U^* \phi_0, \psi\rangle_H dU = \int_{\mathcal{U}} \langle E(T)\phi_0, \psi\rangle_H dU\\
&= \langle E(T)\phi_0, \psi\rangle_H \textstyle\int_{\mathcal{U}} dU = \langle E(T)\phi_0, \psi\rangle_H
\end{aligned}$$

Thus, $E(S_T)\phi_0 = E(T)\phi_0$. Hence $E(S_T) = E(T)$ and

$$I(E(T)) = I(E(S_T)) = S_T = \int_{\mathcal{U}} UTU^* dU$$

Finally, assume that

$$I(E(T)) \notin \overline{\mathrm{conv}}^{\sigma w}\{UTU^* \mid U \in \mathcal{U}\}$$

Then there exist $\omega \in \tilde{M}_*$ and $c, d \in \mathbb{R}$, $c < d$, such that $\mathrm{Re}\,\omega(UTU^*) \geq d$ for $U \in \mathcal{U}$ and $\mathrm{Re}\,\omega(I(E(T))) \leq c$. But:

$$\begin{aligned}
&\mathrm{Re}\,\omega(I(E(T))) = \mathrm{Re}\,\omega(S_T) \\
&= \mathrm{Re}\,\omega\left(\textstyle\int_{\mathcal{U}} UTU^* dU\right) = \mathrm{Re}\,\textstyle\int_{\mathcal{U}} \omega(UTU^*)dU \\
&= \textstyle\int_{\mathcal{U}} \mathrm{Re}\,\omega(UTU^*)dU \geq \textstyle\int_{\mathcal{U}} d \cdot dU = d > c
\end{aligned}$$

This is a contradiction.

□

Lemma 15.9 (See [25, Lemma 2.6].) *Let $\mathcal{S}$ be a σ-weakly closed $I(M)$-bimodule in $\tilde{M}$, V be a partial isometry in $\tilde{M}$ such that $V^*I(M)V \subset I(M)$. Then for every $T \in \mathcal{S}$:*

$$I(E(TV^*))V \in \mathcal{S}$$

Proof. By Lemma 15.8, there exists a net

$$(T_\alpha) \subset \mathrm{conv}\{UTV^*U^* \mid U \in \mathcal{U}\}$$

such that $T_\alpha \overset{\sigma w}{\to} E(TV^*)$. Then $T_\alpha V \overset{\sigma w}{\to} E(TV^*)V$. But for every $U \in \mathcal{U}$:

$$UT\underbrace{V^*U^*V}_{\in I(M)} \in I(M)\,\mathcal{S}\,I(M) \subseteq \mathcal{S}$$

Hence for every α: $T_\alpha V \in \mathcal{S}$. Since $\mathcal{S}$ is σ-weakly closed, $E(TV^*)V \in \mathcal{S}$.

□

Notation. For every $\tau \in \Sigma$ let

$$\mathcal{Q}_{\mathcal{S},\tau} = \{a \in M \mid I(a)\pi(\tau) \in \mathcal{S}\}$$

In what follows, let $\tau \in \Sigma$ be arbitrary.

Lemma 15.10 *$\mathcal{Q}_{\mathcal{S},\tau}$ is a σ-weakly closed two-sided ideal of M.*

Proof.

1. Let $(a_\alpha) \subset \mathcal{Q}_{\mathcal{S},\tau}$, $a_\alpha \overset{\sigma w}{\to} a$. Then $I(a_\alpha)\pi(\tau) \overset{\sigma w}{\to} I(a)\pi(\tau)$. But $\mathcal{S}$ is a σ-weakly closed hence $I(a)\pi(\tau) \in \mathcal{S}$ and $a \in \mathcal{Q}_{\mathcal{S},\tau}$.

2. Let $a \in \mathcal{Q}_{\mathcal{S},\tau}$, $b \in M$.

$$I(ba)\pi(\tau) = I(b)(I(a)\pi(\tau)) \in \mathcal{S}$$

hence $ba \in \mathcal{Q}_{\mathcal{S},\tau}$.

3.
$$I(ab)\pi(\tau) = I(a)I(b)\pi(\tau) = I(a)\pi(\tau) \cdot \pi(\tau)^* I(b)\pi(\tau) \in \mathcal{S}$$

because $I(a)\pi(\tau) \in \mathcal{S}$, $\pi(\tau)I(b)\pi(\tau)^* \in I(M)$. Hence $ab \in \mathcal{Q}_{\mathcal{S},\tau}$.

□

NOTATION. Let $Q_{\mathcal{S},\tau}$ be the maximal projection of the ideal $\mathcal{Q}_{\mathcal{S},\tau}$.
Let $P_{\mathcal{S},\tau} \in Z(M)$ be defined as follows:

$$P_{\mathcal{S},\tau}(x) = \alpha_{(x,\tau x)}(P_{\mathcal{S}}[x, \tau x])$$

for a. e. $x \in X$.

Then

$$(I(P_{\mathcal{S},\tau})\pi(\tau))(x, \tau x) = P_{\mathcal{S}}[x, \tau x]$$

Lemma 15.11 $Q_{\mathcal{S},\tau} = P_{\mathcal{S},\tau}$.

PROOF.

1. $Q_{\mathcal{S},\tau} \in \mathcal{Q}_{\mathcal{S},\tau}$ hence $I(Q_{\mathcal{S},\tau})\pi(\tau) \in \mathcal{S}$. Then by Theorem 15.6, $I(Q_{\mathcal{S},\tau})\pi(\tau) \in \mathcal{T}(P_{\mathcal{S}})$. In particular, for a. e. $x \in X$:

$$\begin{aligned} I(Q_{\mathcal{S},\tau}\pi(\tau))(x, \tau x) &= I(Q_{\mathcal{S},\tau}\pi(\tau))(x, \tau x)P_{\mathcal{S}}[x, \tau x] \\ &= I(Q_{\mathcal{S},\tau}\pi(\tau))(x, \tau x)(I(P_{\mathcal{S},\tau}\pi(\tau))(x, \tau x) \quad (15.1) \end{aligned}$$

But for $a \in M$:

$$(I(a)\pi(\tau))(x, \tau x) = \alpha_{(\tau x, x)}(a(x))$$

hence (15.1) means that

$$\alpha_{(\tau x,x)}(Q_{\mathcal{S},\tau}(x)) = \alpha_{(\tau x,x)}(Q_{\mathcal{S},\tau}(x)) \cdot \alpha_{(\tau x,x)}(P_{\mathcal{S},\tau}(x))$$

or: $Q_{\mathcal{S},\tau}(x) = Q_{\mathcal{S},\tau}(x)P_{\mathcal{S},\tau}(x)$ for a. e. $x \in X$. Thus, $Q_{\mathcal{S},\tau}(x) \leq P_{\mathcal{S},\tau}(x)$ a. e. and $Q_{\mathcal{S},\tau} \leq P_{\mathcal{S},\tau}$.

2. Let $P'_{\mathcal{S}} \in \tilde{Z}^p$ be defined as follows:

$$P'_{\mathcal{S}}[x, y] = \begin{cases} P_{\mathcal{S}}[x, y] & \text{if } y \neq \tau x \\ 0 & \text{if } y = \tau x \end{cases}$$

Then, by part (1) of this proof:

$$P'_{\mathcal{S}}[x, \tau x] = \alpha_{(\tau x,x)}(Q_{\mathcal{S},\tau}(x)) \leq \alpha_{(\tau x,x)}(P_{\mathcal{S},\tau}(x)) = P_{\mathcal{S}}[x, \tau x]$$

Thus, $P'_{\mathcal{S}} \leq P_{\mathcal{S}}$.

Now let $T \in \mathcal{S}$. Let $b = E(T\pi(\tau)^*)$. By Lemma 15.9, $I(b)\pi(\tau) \in \mathcal{S}$, hence $b \in \mathcal{Q}_{\mathcal{S},\tau}$. But

$$\begin{aligned} b(x) &= (T\pi(\tau)^*)(x,x) = (T\pi(\tau^{-1}))(x,x) \\ &= \textstyle\sum_z \alpha_{(x,z)}(T(x,z))c(x,z,x)\pi(\tau^{-1})(z,x)c(z,x,x) \\ &= \alpha_{(x,\tau x)}(T(x,\tau x)) \end{aligned}$$

Thus,

$$\alpha_{(x,\tau x)}(T(x,\tau x)) = b(x) = b(x)Q_{\mathcal{S},\tau}(x) = \alpha_{(x,\tau x)}(T(x,\tau x))Q_{\mathcal{S},\tau}(x)$$

In other words:

$$T(x,\tau x) = T(x,\tau x)\alpha_{(x,\tau x)}(Q_{\mathcal{S},\tau}(x)) = T(x,\tau x)P'_{\mathcal{S}}[x,\tau x]$$

For $y \neq \tau x$ we have: $T \in \mathcal{S} \subseteq \mathcal{T}(P_{\mathcal{S}})$, hence

$$T(x,y) = T(x,y)P_{\mathcal{S}}[x,y] = T(x,y)P'_{\mathcal{S}}[x,y]$$

Thus, $T \in \mathcal{T}(P'_{\mathcal{S}})$.

This is true for every $T \in \mathcal{S}$, hence $\mathcal{S} \subseteq \mathcal{T}(P'_{\mathcal{S}})$, hence $P'_{\mathcal{S}} \in \mathcal{P}_{\mathcal{S}}$, so $P_{\mathcal{S}} \leq P'_{\mathcal{S}}$. In particular, $P_{\mathcal{S}}[x,\tau x] \leq P'_{\mathcal{S}}[x,\tau x]$, hence for a. e. $x \in X$:

$$P_{\mathcal{S},\tau}(x) = \alpha_{(x,\tau x)}(P_{\mathcal{S}}[x,\tau x]) \leq \alpha_{(x,\tau x)}(P'_{\mathcal{S}}[x,\tau x]) = Q_{\mathcal{S},\tau}(x)$$

Thus, $P_{\mathcal{S},\tau} \leq Q_{\mathcal{S},\tau}$.

□

Corollary 15.12 *Let $T = I(a)\pi(\tau)$ for some $a \in M$ and $\tau \in \Sigma$. If $T \in \mathcal{T}(P_{\mathcal{S}})$ then $T \in \mathcal{S}$.*

PROOF. For a. e. $x \in X$:

$$\begin{aligned} a(x) &= \alpha_{(x,\tau x)}(\alpha_{(\tau x,x)}(a(x))) = \alpha_{(x,\tau x)}(T(x,\tau x)) \\ &= \alpha_{(x,\tau x)}(T(x,\tau x)P_{\mathcal{S}}[x,\tau x]) = \alpha_{(x,\tau x)}(T(x,\tau x))\alpha_{(x,\tau x)}(P_{\mathcal{S}}[x,\tau x]) \\ &= \alpha_{(x,\tau x)}(\alpha_{(\tau x,x)}(a(x)))P_{\mathcal{S},\tau}(x) = a(x)P_{\mathcal{S},\tau}(x) = a(x)Q_{\mathcal{S},\tau}(x) \end{aligned}$$

Hence $a = a \cdot Q_{\mathcal{S},\tau}(x)$. Thus, $a \in \mathcal{Q}_{\mathcal{S},\tau}$ so by the definition of $\mathcal{Q}_{\mathcal{S},\tau}$, $T = I(a)\pi(\tau) \in \mathcal{S}$.

□

Corollary 15.13

$$\mathcal{T}(P_{\mathcal{S}}) \cap \tilde{M}_0 \subset \mathcal{S}$$

PROOF. Let $T \in \mathcal{T}(P_{\mathcal{S}}) \cap \tilde{M}_0$. There exist $n \in \mathbb{N}$, $a_i \in M$ and $\tau_i \in \Sigma^p$ for $i = 1,\ldots,n$ (i. e. τ_i are partial Borel bijections) such that the graphs of τ_i are pairwise disjoint and $T = \sum_{i=1}^n I(a_i)\pi(\tau_i)$. Then for every $i = 1,\ldots,n$:

$$(I(a_i)\pi(\tau_i))(x,y) = \begin{cases} T(x,y) & \text{if } x \in \mathcal{D}(\tau_i), \quad y = \tau_i x \\ 0 & \text{otherwise} \end{cases}$$

Hence,

$$(I(a_i)\pi(\tau_i))(x,y) = (I(a_i)\pi(\tau_i))(x,y)P_{\mathcal{S}}[x,y]$$

and $I(a_i)\pi(\tau_i) \in \mathcal{T}(P_{\mathcal{S}})$ for $i = 1,\ldots,n$. By Corollary 15.12, for every $i = 1,\ldots,n$: $I(a_i)\pi(\tau_i) \in \mathcal{S}$, so $T = \sum_{i=1}^n I(a_i)\pi(\tau_i) \in \mathcal{S}$.

□

Lemma 15.14 (See [26, Lemma 4(1)].) *Let R_n be an equivalence relation on a standard Borel measure space $(X_n, \mathcal{B}_n, \mu_n)$ such that every coset of R_n consists just on n elements. Then there exists a Borel measure space $(X', \mathcal{B}', \mu')$ such that the space $(X_n, \mathcal{B}_n, \mu_n)$ is isomorphic to*

$$\left(S_n \times X', 2^{S_n} \times \mathcal{B}', (\text{counting measure}) \times \mu'\right)$$

and R_n is isomorphic to the equivalence relation on $S_n \times X'$ defined by:

$$(s, x') \sim (t, y') \iff x' = y', \qquad \text{where} \quad s, t \in S_n, \quad x', y' \in X'$$

(Here S_n is a set with n elements.)

PROOF. Let Q be the space of cosets of R_n. The Borel structure of X_n can be translated to Q by the canonical surjection $x \mapsto [x]$. (Here $[x]$ is the coset of x.) Let μ' be the image of μ.

The space (X_n, μ_n) is Borel isomorphic to the line segment $[0, 1]$ (see [23, beginning of Section 3]). By [6, the proof of Proposition 3.2.72, part (2)], there exists a Borel section $f_1 : Q \to X_n$ such that $[f_1(q)] = q$ for every $q \in Q$. By [6, the proof of Proposition 3.2.72, part (1)], f_1 is a Borel isomorphism onto its image. In particular, the set $F_1 = \text{Im}(f_1)$ is Borel.

Let $X_{n-1} = X_n \setminus F_1$. Let $R_{n-1} = R_n \cup (X_{n-1} \times X_{n-1})$. Then R_{n-1} is an equivalence relation, and every coset of R_{n-1} consists just on $n - 1$ elements. The space of cosets of R_{n-1} is Q yet.

By the same process we can find a Borel set $F_2 \subset X_{n-1}$ which is isomorphic to Q (and to F_1).

Continuing this process we get:

1. $X_n = F_1 \cup F_2 \cup \cdots \cup F_n$;

2. for every $i = 1, 2, \ldots, n$: F_i is isomorphic to Q;

3. for every $x, y \in X_n$:

$$x \sim y \iff [x] = [y] \iff x = f_i(q), \quad y = f_j(q)$$

for some $q \in Q$, $1 \leq i \leq n$, $1 \leq j \leq n$.

Now, let every $x = f_i(q)$ correspond to the pair (i, q). This correspondence is just what the lemma asserts.

□

Corollary 15.15 *If every coset of R consists just on n elements, then $\tilde{M} = \tilde{M}_0$.*

PROOF. (See [14, Proposition 4.1].) It is enough to represent R as finite union of graphs of partial isomorphisms. By Lemma 15.14, we can assume that $X = S_n \times X'$, $\text{card}(S_n) = n$, $\mu' = (\text{counting measure}) \times \mu'$, and $(s, x') \overset{R}{\sim} (t, y')$ if and only if $x' = y'$. Let $\tau : X \to X$ be defined as

$$\tau((s, x')) = ((s + 1) \bmod n, x')$$

Then τ is a partial Borel homomorphism, and

$$R = \bigcup_{i=1}^{n} \Gamma(\tau_i)$$

□

Corollary 15.16 *If cardinalities of all cosets of R are equal, then for every σ-weakly closed bimodule $\mathcal{S}$:*

$$\mathcal{S} = \mathcal{T}(P_{\mathcal{S}})$$

Theorem 15.17 *If R is finite (i. e. every coset of R is finite), then for every σ-weakly closed bimodule $\mathcal{S}$:*

$$\mathcal{S} = \mathcal{T}(P_{\mathcal{S}})$$

PROOF.

1. For every $n \in \mathbb{N}$, let $X_n = \{x \in X \mid \text{card}([x]) = n\}$. By [14, Theorem 2(a)], all X_n are Borel. Let $R_n = R \cap (X_n \times X_n)$, $n \in \mathbb{N}$. Then every R_n has cosets consisting just on n enements, and $R = \bigcup_{n=1}^{\infty} R_n$. Let

$$\begin{aligned} H_n &= \textstyle\int_{X_n} H(x)d\mu(x), & \tilde{H}_n &= \textstyle\int_{R_n} H(x,y)d\nu(x,y), \\ M_n &= \textstyle\int_{X_n} M(x)d\mu(x), & \tilde{M}_n &= M_n \bowtie R_n, \\ \tilde{Z}_n &= \textstyle\int_{R_n} Z(M((x,y))d\nu(x,y) \end{aligned}$$

Then $H = \sum_n^{\oplus} H_n$, $\tilde{H} = \sum_n^{\oplus} \tilde{H}_n$, $\tilde{Z} = \sum_n^{\oplus} \tilde{Z}_n$. It is clear that $\tilde{M}_n \cong I(\chi_{X_n})\tilde{M}$, hence $\tilde{M} = \sum_n^{\oplus} \tilde{M}_n$.

2. Let for every σ-wealy closed bimodule $\mathcal{T}$:

$$\mathcal{T}_n = \{T_n \mid \text{there exists } T \in \mathcal{T} : T = \sum_{n=1}^{\infty} T_n \text{ is the decomposition of the operator } T\}$$

Then $\mathcal{T} = \sum_{n=1}^{\infty} \mathcal{T}_n$. Indeed, "$\subset$" is clear, and "$\supset$" follows because $\mathcal{T}_n \cong I(\chi_{X_n})\mathcal{T}$.

Furthermore, $\mathcal{T}_n$ is a σ-weakly closed bimodule in $\tilde{M}_n$.

3. Let $P \in \tilde{Z}$, $P = \sum_n P_n$ be the decomposition of P, $n \in \mathbb{N}$. Then $\mathcal{T}(P_n) = \mathcal{T}(P)_n$. Indeed, $\tilde{M}_n = \mathcal{T}(\chi_{R_n})$, hence

$$\begin{aligned} &\mathcal{T}(P)_n = \mathcal{T}(P) \cap \tilde{M}_n = \mathcal{T}(P) \cap \mathcal{T}(\chi_{R_n}) \\ &= \mathcal{T}(P \cdot \chi_{R_n}) = \mathcal{T}(P_n) \end{aligned}$$

(Here $\tilde{M}_n$ are assumed to be embedded into $\tilde{M}$.)

4. It follows from the definitions that

$$\mathcal{S} \subseteq \mathcal{T}(P) \iff \text{for every } n \in \mathbb{N} : \mathcal{S}_n \subseteq \mathcal{T}(P_n)$$

Hence

$$(P_{\mathcal{S}})_n = \Big(\inf\{P \mid \mathcal{S} \subseteq \mathcal{T}(P)\}\Big)_n = \inf\{P_n \mid \mathcal{S} \subseteq \mathcal{T}(P)\}$$
$$= \inf\{P_n \mid \mathcal{S}_n \subseteq \mathcal{T}(P_n)\} = P_{\mathcal{S}_n}$$

5. Finally, by Corollary 15.16, for every $n \in \mathbb{N}$: $\mathcal{S}_n = \mathcal{T}(P_{\mathcal{S}_n})$. Thus,

$$\mathcal{S} = \sum \mathcal{S}_n = \sum \mathcal{T}(P_{\mathcal{S}_n}) = \sum \mathcal{T}((P_{\mathcal{S}})_n)$$
$$= \sum (\mathcal{T}(P_{\mathcal{S}}))_n = \mathcal{T}(P_{\mathcal{S}})$$

□

Theorem 15.18 (Spectral Theorem for Bimodules) *Let R be hyperfinite. Then for every σ-weakly closed bimodule $\mathcal{S}$:*

$$\mathcal{S} = \mathcal{T}(P_{\mathcal{S}})$$

PROOF.

1. Let R be hyperfinite. Then $R = \bigcup_{n=1}^{\infty} R_n$, where every R_n is a finite equivalence relation on X, and for $n \in \mathbb{N}$: $R_n \subset R_{n+1}$.

   For every $n \in \mathbb{N}$, $\chi_{R_n} \in \tilde{Z}$. (Here χ_{R_n} is the characteristic function of R_n identified with the operator of multiplication on this function.) It is clear that

$$\chi_{R_n} \circ \chi_{R_n} = \chi_{R_n} = \theta(\chi_{R_n})$$

   and $\chi_{R_n}(x,x) = 1_{H(x)}$ for $x \in X$. By Theorem 14.4, $\mathcal{T}(\chi_{R_n})$ is a von Neumann subalgebra of $\tilde{M}$, and there exists a conditional expectation Φ_n from $\tilde{M}$ onto $\mathcal{T}(\chi_{R_n})$, which is given by the following formula:

$$\Phi(T)(x,y) = \chi_{R_n}[x,y]T(x,y)$$

2. For every $n \in \mathbb{N}$, let $\tilde{H}_n = \int_{R_n} H(x,y)d\nu(x,y)$. Identifying $\tilde{H}_n$ as a subspace of $\tilde{H}$ we have:

$$\mathcal{T}(\chi_{R_n}) = M \bowtie R_n$$

   Indeed, "$\supseteq$" is clear. To prove "$\subseteq$" we note that:

   (a) $\mathcal{T}(\chi_{R_n}) \subseteq B(\tilde{H}_n)$;

   (b) elements of $\mathcal{T}(\chi_{R_n})$ commute with $I'(M')$ and with $\pi'(\tau)$ for $\tau \in \Sigma$ such that $\Gamma(\tau) \subset R_n$.

3. For a σ-weakly closed bimodule $\mathcal{T}$ in $\tilde{M}$ and for $n \in \mathbb{N}$ write: $\mathcal{T}_n = \Phi_n(T)$. Then $\mathcal{T}_n$ is a σ-weakly closed bimodule in $\tilde{M}_n$. Let for $P \in \tilde{Z}^p$, $n \in \mathbb{N}$:

$$P_n[x, y] = \chi_{R_n}[x, y]P[x, y]$$

Then $P_n \in \tilde{Z}_n^p$, where

$$\tilde{Z}_n = \int_{R_n} Z(M(x, y))d\nu(x, y)$$

and $\tilde{Z}_n$ is embedded into $\tilde{Z}$.

It is clear that $(\mathcal{T}(P))_n = \mathcal{T}(P_n)$.

4. Let $\mathcal{S}$ be a σ-weakly closed bimodule in $\tilde{M}$. With the notation above:

$$(P_{\mathcal{S}})_n = P_{(\mathcal{S}_n)}$$

Indeed,

$$\begin{aligned}(P_{\mathcal{S}})_n &= \left(\inf\{P \in \tilde{Z}^p \mid \mathcal{S} \subseteq \mathcal{T}(P)\}\right)_n \\ &= \inf\{P_n \mid P \in \tilde{Z}^p,\ \mathcal{S} \subseteq \mathcal{T}(P)\}\end{aligned}$$

and

$$P_{\mathcal{S}_n} = \inf\{P \in \tilde{Z}^p \mid \mathcal{S}_n \subseteq \mathcal{T}(P)\}$$

By the definitions, if $\mathcal{S} \subseteq \mathcal{T}(P)$ then $\mathcal{S}_n \subseteq \mathcal{T}(P_n)$ hence in particular $\mathcal{S}_n \subseteq \mathcal{T}((P_{\mathcal{S}})_n)$ hence $P_{\mathcal{S}_n} \leq (P_{\mathcal{S}})_n$.

Conversely, if $\mathcal{S}_n \subseteq \mathcal{T}(P)$ then $\mathcal{S} \subseteq \mathcal{T}(Q)$ where $Q = \sup(P, 1_{\tilde{H}_n} - \chi_{R_n})$, hence $P_{\mathcal{S}} \leq Q$ and $(P_{\mathcal{S}})_n \leq Q_n = P_n$. In particular, $(P_{\mathcal{S}})_n \leq (P_{\mathcal{S}_n})_n = P_{\mathcal{S}_n}$.

5. Let $T \in \mathcal{T}(P_{\mathcal{S}})$, $n \in \mathbb{N}$.

$$\Phi_n(T) \in (\mathcal{T}(P_{\mathcal{S}}))_n = \mathcal{T}((P_{\mathcal{S}})_n) = \mathcal{T}(P_{\mathcal{S}_n}) = \mathcal{S}_n \subseteq \mathcal{S}$$

All Φ_n are conditional expectations, hence $\|\Phi(T)\| \leq \|T\|$ for every $n \in \mathbb{N}$.

6. R_n are increasing, and $R = \bigcup_{n=1}^{\infty} R_n$. Hence $\chi_{R_n} \to 1_{\tilde{H}}$ strongly (as elements of $B(\tilde{H})$).

$$\begin{aligned}(\Phi_n(T)\tilde{\phi}_0)(x, y) &= \Phi_n(T)(x, y)c(x, y, y)\tilde{\phi}_0(y, y) \\ &= \chi_{R_n}[x, y]T(x, y)c(x, y, y)\tilde{\phi}_0(y, y) \\ &= \chi_{R_n}[x, y](T\tilde{\phi}_0)(x, y) = (\chi_{R_n}T\tilde{\phi}_0)(x, y)\end{aligned}$$

Thus, $\Phi_n(T)\tilde{\phi}_0 = \chi_{R_n}T\tilde{\phi}_0 \to T\tilde{\phi}_0$. Hence, for $S' \in \tilde{M}'$:

$$\Phi_n(T)S'\tilde{\phi}_0 = S'\Phi_n(T)\tilde{\phi}_0 \to S'T\tilde{\phi}_0 = TS'\tilde{\phi}_0$$

Finally, for every $\psi \in \tilde{H}$, $\varepsilon > 0$ there exists $S'_\varepsilon \in \tilde{M}'$ such that $\|S'_\varepsilon\tilde{\phi}_0 - \psi\| \leq \varepsilon/(3\|T\|)$. Let $N \in \mathbb{N}$ be such that for $n \geq N$:

$$\|\Phi_n(T)S'_\varepsilon\tilde{\phi}_0 - TS'_\varepsilon\tilde{\phi}_0\| < \frac{\varepsilon}{3}$$

Then for $n \geq N$:

$$\begin{aligned}
&\|\Phi_n(T)\psi - T\psi\| \\
&\leq \|\Phi_n(T)\psi - \Phi_n(T)S'_\varepsilon\tilde{\phi}_0\| + \|\Phi_n(T)S'_\varepsilon\tilde{\phi}_0 - TS'_\varepsilon\tilde{\phi}_0\| + \|TS'_\varepsilon\tilde{\phi}_0 - T\psi\| \\
&\leq \|\Phi_n(T)\| \cdot \|\psi - S'_\varepsilon\tilde{\phi}_0\| + \|\Phi_n(T)S'_\varepsilon\tilde{\phi}_0 - TS'_\varepsilon\tilde{\phi}_0\| + \|T\| \cdot \|S'_\varepsilon\tilde{\phi}_0 - \psi\| \\
&< \|T\| \cdot \frac{\varepsilon}{3\|T\|} + \frac{\varepsilon}{3} + \|T\| \cdot \frac{\varepsilon}{3\|T\|} = \varepsilon
\end{aligned}$$

Thus, $\Phi_n(T)\psi \to T\psi$, hence $\Phi_n(T) \to T$ strongly (and σ-strongly). Hence, $T \in \mathcal{S}$.

□

CHAPTER 16

ANALYTIC ALGEBRA OF A FLOW OF AUTOMORPHISMS

Definition 16.1 (See [2, Definition 2.1] and [22, Definition 2.17].) *A* **flow of isomorphisms** *of the algebra* $\tilde{M}$ *is a* σ*-weakly continuous representation* $\{\gamma_t\}_{t\in\mathbb{R}}$ *of* $\mathbb{R}$ *as automorphisms of* $\tilde{M}$.

The **Arveson spectrum** *of an element* $x \in \tilde{M}$ *with respect to a flow* γ *is the subset* $\operatorname{sp}_\gamma(x)$ *of* $\mathbb{R}$ *defined as follows:*

$$\operatorname{sp}_\gamma(x) = \Big\{p \in \mathbb{R} \mid \text{ for every } f \in L^1(\mathbb{R}),$$
$$\textstyle\int_{\mathbb{R}} f(t)\gamma_t(x)dt = 0 \Longrightarrow \int_{\mathbb{R}} f(t)\exp(-2\pi ipt)dt = 0\Big\}$$

The **analytic algebra** *of the flow* γ *is the algebra* $H^\infty(\gamma)$ *defined as follows:*

$$H^\infty(\gamma) = \big\{x \in \tilde{M} \mid \operatorname{sp}_\gamma(x) \subseteq [0, +\infty)\ \big\}$$

NOTATION. Let E be the self-adjoint operator affiliated to $\tilde{Z}$ which is associated with γ by Theorem 11.4. Let for $s \in \mathbb{R}$:

$$E_s = \chi_{[s,+\infty)}(E)$$

Then $E_s[x, y] = \chi_{[s,+\infty)}(E[x, y])$.

Lemma 16.2 *Let* E *be a self-adjoint (not necessarily bounded) operator,* U *be a unitary operator. Then:*

1. UEU^* *is self-adjoint;*
2. *for every* $s \in \mathbb{R}$*:*

$$\chi_{[s,+\infty)}(UEU^*) = U\chi_{[s,+\infty)}(E)U^*$$

PROOF.

1. It is easy to see that UE is closed.
2. $(UEU^*)^* = (EU^*)^*U^* = (UE)^{**}U^* = \overline{UE}U^* = UEU^*$.
3. Let

$$\begin{aligned} H_1 &= \operatorname{Im}\big(\chi_{[s,+\infty)}(UEU^*)\big), \\ H_2 &= \operatorname{Im}\big(U\chi_{[s,+\infty)}(E)U^*\big) \end{aligned}$$

Then

$$\begin{aligned} &\xi \in H_1 \iff \langle UEU^*\xi, \xi\rangle \geq s \cdot \|\xi\|^2 \\ &\iff \langle EU^*\xi, U^*\xi\rangle \geq s \cdot \|U^*\xi\|^2 \iff U^*\xi \in \operatorname{Im}(\chi_{[s,+\infty)}(E)) \\ &\iff \xi \in \operatorname{Im}(U\chi_{[s,+\infty)}(E)) = \operatorname{Im}(U\chi_{[s,+\infty)}(E)U^*) = H_2 \end{aligned}$$

□

Lemma 16.3 *For $s \in \mathbb{R}$, a. e. $x \sim y \sim z$:*

$$E_s[x,y] \geq \alpha_{(y,z)}(E_0[x,z])c(x,z,y)E_s[z,y]c(x,z,y)^{-1}$$

PROOF. The operator E above satisfies the following property (see (11.1)):

$$E(z,y) = \alpha_{(y,x)}(E(z,x)) + c(z,x,y)E(x,y)c(z,x,y)^{-1}$$

Now, for two commuting self-adjoint operators a and b:

$$\chi_{[s,+\infty)}(a+b) \geq \chi_{[0,+\infty)}(a)\cdot\chi_{[s,+\infty)}(b)$$

(This is true because this is true for functions: if $a(\omega) \geq 0$ and $b(\omega) \geq s$ then $(a+b)(\omega) \geq s$.)

Let $a = \alpha_{(y,x)}(E[z,x])$ and $b = c(z,x,y)E[x,y]c(z,x,y)^{-1}$. Then $a+b = E[z,y]$, and:

$$\begin{aligned}
E_s[x,y] &= \chi_{[s,+\infty)}(E[x,y]) = \chi_{[s,+\infty)}(a+b)\\
&\geq \chi_{[0,+\infty)}(a)\cdot\chi_{[s,+\infty)}(b)\\
&= \chi_{[0,+\infty)}(\alpha_{(y,x)}(E[z,x]))\cdot\chi_{[s,+\infty)}(c(z,x,y)E[x,y]c((z,x,y)^{-1})\\
&= \alpha_{(y,x)}(\chi_{[0,+\infty)}(E[z,x]))\cdot c(z,x,y)\chi_{[s,+\infty)}(E[x,y])c(z,x,y)^{-1}\\
&= \alpha_{(y,x)}(E_0[z,x])\cdot c(z,x,y)E_s[x,y]c(z,x,y)^{-1}
\end{aligned}$$

□

Theorem 16.4 *Let γ be a flow of automorphisms of $\tilde{M}$ leaving $I(M)$ fixed pointwise. Let $E\eta\tilde{Z}$ be the self-adjoint operator associated with γ by Theorem 11.4. Then*

$$H^\infty(\gamma) = \mathcal{T}(E_0)$$

where $E_0 = \chi_{[0,+\infty)}(E)$.

PROOF. By Theorem 11.4, the automorphisms γ_t are of the following kind:

$$\gamma_t(T) = \exp(itE)T\exp(-itE), \qquad T \in \tilde{M}, \quad t \in \mathbb{R}$$

It is known (see [22, Theorem 2.9]) that

$$H^\infty(\gamma) = \{T \in \tilde{M} \mid TE_s = E_sTE_s \text{ for every } s \in \mathbb{R}\}$$

First, let $T \in H^\infty(\gamma)$. Then, in particular:

$$TE_0 = E_0TE_0$$

Moreover, for a. e. $y \in X$: $E[y,y] = 0$, so $E_0[y,y] = 1_{H(y)}$.

It is easy to see that $T(x,y)E_0[x,y] = T((x,y)$ a. e. , hence $T \in \mathcal{T}(E_0)$.

Conversely, let $T \in \mathcal{T}(E_0)$, so that for a. e. $(x,y) \in R$:
$T(x,y) = E_0[x,y]T(x,y)$. Let $s \in \mathbb{R}$, $g \in \tilde{H}$. The straight calculation shows that $E_sTE_sg = TE_sg$.

□

CHAPTER 17

PROPERTIES OF $\tilde{M}$

NOTATION. For every $\tau \in \Sigma$, let ρ_τ be the automorphism of M defined by the following:

$$I(\rho_\tau(a)) = \pi(\tau)I(a)\pi(\tau)^{-1}$$

It is clear that the restriction of ρ_τ onto $Z(M)$ is an automorphism of $Z(M)$.

Definition 17.1 *A trace on M is called* ***R*-invariant**, *if it is invariant with respect to every ρ_τ for $\tau \in \Sigma$. The pair (R, M) is said to be:*

- **of semifinite type**, *if there exists a faithful normal semifinite (f. n. s.) R-invariant trace on M;*
- **of finite type**, *if there exists a faithful normal finite (f. n. f.) R-invariant trace on M;*
- **of purely infinite type**, *if there no exists any f. n. s. R-invariant trace on M;*
- **of properly infinite type**, *if there no exists any f. n. f. R-invariant trace on M.*

Lemma 17.2 *For $T \in \tilde{M}$, $\tau \in \Sigma$:*

$$I(E(\pi(\tau)T\pi(\tau)^{-1})) = \pi(\tau)I(E(T))\pi(\tau)^{-1}$$

PROOF. The proof is straightforward.

□

Corollary 17.3

$$E(\pi(\tau)T\pi(\tau)^{-1}) = E(\pi(\tau)I(E(T))\pi(\tau)^{-1})$$

PROOF. $E \circ I = \mathrm{id}|M$.

□

Theorem 17.4 *The pair (R, M) is of semifinite (resp. finite, purely infinite, properly infinite) type if and only if $\tilde{M}$ is of semifinite (resp. finite, purely infinite, properly infinite) type.*

PROOF. If ϕ is a f. n. s. (resp. f. n. f.) R-invariant trace on M, we define

$$\tilde{\phi}(T) = \phi(E(T)), \qquad T \in \tilde{M}^+$$

Then $\tilde{\phi}$ is a f. n. s. (recp. f. n. f.) weight on $\tilde{M}$. It is easy to see that $\tilde{\phi}$ is a trace.
Conversely, let $\tilde{\phi}$ be a f. n. s. (resp. f. n. f.) trace on $\tilde{M}$. We define:

$$\phi(a) = \tilde{\phi}(I(a))$$

Then ϕ is a f. n. s. (resp. f. n. f.) trace on M. Again, it it easy to see that ϕ is R-invariant.

□

Theorem 17.5 *$\tilde{M}$ is a factor if and only if there are no non-scalar operators in $Z(M)$ which are invaruiant with respect to all of ρ_τ for $\tau \in \Sigma$.*

PROOF. If some operator $a \in Z(M)$ is invariant with respect to all of ρ_τ, then for every $\tau \in \Sigma$:

$$\pi(\tau)I(a)\pi(\tau)^* = I(a)$$

Hence $I(a)$ commutes with every $\pi(\tau)$ for $\tau \in \Sigma$, hence $I(a) \in \tilde{M}'$, hence $I(a) \in Z(\tilde{M})$.
Conservely, let $T \in Z(\tilde{M})$. Then in particular $T \in I(A)' \cap \tilde{M} = I(M)$. Let $T = I(a)$, $a \in M$. But T commutes with $I(M)$, hence $a \in Z(M)$. Finally, T commutes with every $\pi(\tau)$ with $\tau \in \Sigma$, hence a is R-invariant.

□

Corollary 17.6 *If $\tilde{M}$ is a factor then R is ergodic.*

PROOF. If $E \subset X$ is an invariant set with respect to R, then $\chi_E \in Z(M)$. It is easy to see that χ_E is ρ_τ-invariant.

□

REMARK. If $Z(M) = A$ (i. e. the decomposition $M = \int_X M(x)d\mu(x)$ is central), then the converse is true too: if R is ergodic, then $\tilde{M}$ is a factor.
Indeed, let $c \in Z(\tilde{M})$. Then, as in the proof of Theorem 17.4, $c \in I(Z(M)) = I(A)$. If in addition c is a projection that $c = \chi_E$ for some Borel subset $E \subset X$. But c commutes with every $\pi(\tau)$ for $\tau \in \Sigma$. It follows from this that E is τ-invariant for every $\tau \in \Sigma$ hence E is R-invariant.

CHAPTER 18

HYPERFINITENESS AND DILATIONS

We assume that (X, μ) is a standard measure space. By [10, Ch. II, §1, n. 6, Proposition 6, Corollaire], H is separable. Hence, $\tilde{H}$ is separable too.

Lemma 18.1 *Let B be a von Neumann algebra acting on a Hilbert space K and possessing a separating vector. If K is separable, that B_* is separable in the weak topology on B_*.*

PROOF. By [10, Ch. III, §1, n. 4, Théorème 4], for every $\phi \in B_*^+$ there exists $\xi_\phi \in K$ such that $\phi = \omega_{\xi_\phi}$, i. e. $\phi(T) = \langle T\xi_\phi, \xi_\phi \rangle$ for $T \in B$.

Let K_0 be a countable dense subset of K. For some fixed $\phi \in B_*^+$ let $(\xi_i)_{i=1}^\infty \subset K_0$, $\xi_i \to \xi_\phi$. Then $(\xi_i)_{i=1}^\infty$ is a bounded sequence. Let $\|\xi_i\| \leq A_\phi$ for every $i \in \mathbb{R}$. Then for every $T \in B$:

$$\begin{aligned}
&|\omega_{\xi_i}(T) - \phi(T)| = |\langle T\xi_i, \xi_i \rangle - \langle T\xi_\phi, \xi_\phi \rangle| \\
&= |\langle T\xi_i, \xi_i \rangle - \langle T\xi_i, \xi_\phi \rangle + \langle T\xi_i, \xi_\phi \rangle - \langle T\xi_\phi, \xi_\phi \rangle| \\
&\leq |\langle T\xi_i, \xi_i \rangle - \langle T\xi_i, \xi_\phi \rangle| + |\langle T\xi_i, \xi_\phi \rangle - \langle T\xi_\phi, \xi_\phi \rangle| \\
&= |\langle T\xi_i, \xi_i - \xi_\phi \rangle| + |\langle T(\xi_i - \xi_\phi), \xi_\phi \rangle| \\
&\leq \|T\xi_i\| \cdot \|\xi_i - \xi_\phi\| + \|T(\xi_i - \xi_\phi)\| \cdot \|\xi_\phi\| \\
&\leq \|T\| \cdot \|\xi_i\| \cdot \|\xi_i - \xi_\phi\| + \|T\| \cdot \|\xi_i - \xi_\phi\| \cdot \|\xi_\phi\| \\
&\leq \|T\| \cdot A_\phi \cdot \|\xi_i - \xi_\phi\| + \|T\| \cdot \|\xi_i - \xi_\phi\| \cdot \|\xi_\phi\|
\end{aligned}$$

— converges to 0. Thus, the set $\{\omega_\xi \mid \xi \in K_0\}$ is dense in B_*^+, hence the set

$$\{\omega_{\xi_1} - \omega_{\xi_2} + i\omega_{\xi_3} - i\omega_{\xi_4} \mid \xi_i \in K_0,\ k = 1, \ldots, 4\}$$

is dense in B_*.

□

Corollary 18.2 *M_* and $\tilde{M}_*$ are separable.*

Definition 18.3 See [32, Subsections 10.22 and 10.26] and [6, end of Subsection 2.7.3].) *A von Neumann algebra is called* **hyperfinite**, *if it is generated by an increasing sequence of finite dimensional factors.*

A von Neumann algebra is called **σ-approximately finite dimensional** *(**σ-a. f. d.**), if it is generated by an increasing sequence of finite dimensional algebras.*

A von Neumann algebra acting on a Hilbert space K is called **injective**, *if it is a range of some conditional expectation defined on $B(K)$.*

By [29, Theorem XII] and [13, Theorem 3], a factor is hyperfinite if and only if it is σ-a. f. d. (see also [6, end of Subsection 2.7.3]).

By [12, Theorem 2 and Corollary 5], an algebra possessing a separable predual is σ-a. f. d. if and only if it is injective (see also [32, Subsection 10.26]).

Theorem 18.4 *1. If M is σ-a. f. d. and R is hyperfinite then $\tilde{M}$ is σ-a. f. d. .*
2. If $\tilde{M}$ is σ-a. f. d. then M is σ-a. f. d. .

PROOF.

1. Let M be σ-a. f. d. and R be hyperfinite. By [14, Proposition 4.1(b)], there exists an isomorphism $\phi \in \Sigma$ such that R is the relation generated by powers of ϕ. Let $U = \pi(\phi)$. Then $\tilde{M}$ is the von Neumann algebra generated by $I(M)$ and U^n with $n \in \mathbb{Z}$. By [7, Proposition 6.8], $\tilde{M}$ is injective. Hence $\tilde{M}$ is σ-a. f. d.

2. Let $\tilde{M}$ be σ-a. f. d. Then $\tilde{M}$ is injective. Let $F : B(\tilde{H}) \to \tilde{M}$ be a conditional expectation onto $\tilde{M}$. There exists a conditional expectation E from $\tilde{M}$ onto $I(M)$. Then $E \circ F$ is a projector of norm one, hence a conditional expectation, from $B(\tilde{H})$ onto $IM)$. Thus, $I(M)$ is injective, so it is σ-a. f. d. . Hence M is σ-a. f. d. .

□

NOTATION. Let for $f \in L^\infty(R, \nu)$, $\phi \in \Sigma$:

$$f^\phi(x, y) = f(\phi x, y)$$

Let for $g \in A \cong L^\infty(X, \mu)$, $\phi \in \Sigma$:

$$g^\phi(x) = g(\phi x)$$

Definition 18.5 *The center $Z(M)$ of the algebra M will be called* **Σ-amenable**, *if there exists a conditional expectation K from $Z(M)$ onto A such that for every $\phi \in \Sigma$:*

$$(K \circ \rho_\phi)(f) = (K(f))^\phi, \qquad f \in Z(M)$$

REMARK. If $Z(M) = A$ then $Z(M)$ is Σ-amenable with $K = \mathrm{id}_A$.

Lemma 18.6 *For $f \in L^\infty(R, \nu)$, $g \in A$, $\phi \in \Sigma$:*

$$f^\phi = \pi(\phi) f \pi(\phi)^*, \qquad I(g^\phi) = \pi(\phi) I(g) \pi(\phi)^*$$

PROOF. The proof is straightforward.

□

Theorem 18.7 (See [9, Propositions 7 and 10].) *If $\tilde{M}$ is σ-a. f. d. and $Z(M)$ is Σ-amenable, that R is hyperfinite.*

PROOF. Let $\tilde{M}$ be σ-a. f. d. Then $\tilde{M}$ is injective. Let $F : B(\tilde{H}) \to \tilde{M}$ be a conditional expectation onto $\tilde{M}$.

For $f \in L^\infty(R,\nu)$, $a \in M$ we have: $I(a)$ is decomposable with respect to the decomposition $\tilde{H} = \int_R H(x,y)d\nu(x,y)$, hence $I(a)$ commutes with f, and:

$$F(f)I(a) = F(fI(a)) = F(I(a)f) = I(a)F(f)$$

Thus, $F(f) \in I(M)'$. In particular, $F(f) \in I(A)' \cap \tilde{M} = I(M)$. Hence, $F(f) \in I(M) \cap I(M)' = I(Z(M))$. Let $g \in Z(M)$ be such that $F(f) = I(g)$.

Let $k = K(g) = (K \circ I^{-1} \circ F)(f)$, where K is from the definition of Σ-amenability of $Z(M)$. Then $k \in A$, and for every $\phi \in \Sigma$:

$$(K \circ I^{-1} \circ F)(f^\phi) = (K \circ I^{-1} \circ F)(\pi(\phi) f \pi(\phi)^*)$$
$$= (K \circ I^{-1})(\pi(\phi)F(f)\pi(\phi)^*) = (K \circ I^{-1})(\pi(\phi)I(g)\pi(\phi)^*)$$
$$= (K \circ \rho_\phi)(g) = \big(K(g)\big)^\phi = \big((K \circ I^{-1} \circ F)(f)\big)^\phi$$

Thus, $K \circ I^{-1} \circ F$ is a left invariant mean on R. By [9, Proposition 10], R is hyperfinite.

□

Definition 18.8 (See [26, Section 1].) *A σ-weakly closed subalgebra C of a von Neumann algebra B is called a* **σ-Dirichlet subalgebra** *if $C + C^*$ is σ-weakly dense in B.*

A pair (B, C) consisting from a von Neumann algebra B and its σ-Dirichlet subalgebra B will be called **dilable** *if every σ-weakly continuous contraction ρ of C has a dilation to a normal *-representation of B.*

An representation ρ of an algebra B is called **universally completely contractive,** *if for every $n \in \mathbb{N}$, the representation $\rho \otimes I_n$ of the algebra $\mathrm{Mat}_n \otimes B$ is contractive. (Here Mat_n is the algebra of matrices $n \times n$, and I_n is the usual representation of this algebra on n-dimensional space.)*

An algebra C as above will be called **universally completely contractive,** *if every σ-weakly continuous contraction of ρ of C is universally completely contractive.*

REMARK. It is known (see [26, Section 1]) that if ρ as above is universally completely contractive, then ρ has a dilation. Thus, if C is universally completely contractive, then (B, C) is dilable.

In this section we will assume that $c(x, y, z) = 1$ for every $x \sim y \sim z$.

From now on, for every function f on $S \times \{1, \ldots, m\}$ we will write $f(z, i)$ in place of $f((z, i))$.

Lemma 18.9 *Let, in the notation above,* $X = S \times \{1, \ldots, m\}$, $R = \Delta_S \times \{1, \ldots, m\}^2$, $P_S = \chi_{S \times \{1\}}$, $M_S = M_{P_S}$ *(the induced von Neumann algebra).*

Let $\{Q_n\}_{n=1}^{\infty}$ *be an orthogonal family of projections in the center* $Z(M_S)$ *of* M_S *with sum 1.*

Let $\mathcal{T}$ *be a* σ*-Dirichlet subalgebra of* $\tilde{M}$ *containing* $I(M)$, ρ *be a* σ*-weakly continuous contractive representation of* $\mathcal{T}$*. Let* ρ_n *be defined as follows:*

$$\rho(T) = \rho(\bar{Q}_n T \bar{Q}_n), \qquad T \in \mathcal{T}$$

where $\bar{Q}_n \in Z(M)$, $\bar{Q}_n(s,i) = \alpha_{((s,i),(s,1))}(Q_n(z,1))$, $i = 1, \ldots, n$.

Then ρ *is universally completely contractive if and only if every* ρ_n *is universally completely contractive.*

PROOF. Let $P_n = \rho(\bar{Q}_n)$. Then P_n are orthogonal with sum 1, and P_n commute with elements of $\rho(\mathcal{T})$. Moreover,

$$\rho_n(b) = P_n \rho(b) P_n, \qquad b \in \mathcal{T}$$

By [26, Lemma 2], ρ is universally completely contractive if and only if every ρ_n is universally completely contractive.

□

Lemma 18.10 (See [30, Proposition 1.1].) *Let, in the notation above,* $X = \{1, \ldots, n\}$, $R = X \times X$*. Let* $B \subset R$, $\mathcal{T} = \mathcal{T}(\chi_B)$*. Assume in addition that* $\mathcal{T}$ *is a* σ*-Dirichlet subalgebra of* $\tilde{M}$ *containing* $I(M)$*.*

Then every σ*-weakly continuous contractive representation of* $\mathcal{T}$ *has a dilation to a *-representation of* $\tilde{M}$*.*

PROOF. It follows from the definitions that B contains some total order on X. We can assume that B <u>is</u> a total order and that $\mathcal{T}$ is the algebra of all "upper triangular matrices".

Let ρ be a contractive σ-weakly continuous representation of $\mathcal{T}$ on some Hilbert space K. We can assume that $\rho(1_{\tilde{H}}) = 1_K$.

Let $E_{ij} \in \mathcal{T}$ be defined as follows:

$$E_{ij}(x,y) = \delta_{i,x} \cdot \delta_{j,y} \cdot 1_{H(y)}$$

Let $P_i = \rho(E_{ii})$, $i = 1, \ldots, m$. Then P_i are projections. Let $K_i = P_i K$. Then $K = \sum^{\oplus} K_i$. Let $T_{i,j} = P_i \rho(E_{ij}) | K_j$, $i,j = 1, \ldots, m$. Then: $\|T_{i,j}\| \leq 1$, $T_{i,j} : K_j \to K_i$, $T_{i,j} = T_{i,i+1} \cdot T_{i+1,i+2} \cdots T_{j-1,j}$.

Let $T_i = T_{i,i+1}$, $i = 1, \ldots, m-1$.

Let $L_i = K_i \oplus K_i \oplus K_i \oplus \cdots$. Let V_i be the operator acting on L_i $(1 \leq i \leq m-1)$ by following:

$$V_i(k_1, k_2, k_3, \ldots) = (T_i k_1, (1_{K_i} - T_i^* T_i)^{1/2} k_1, k_2, k_3, \ldots)$$

Then V_i are isometries, and

$$T_i T_{i+1} \cdots T_j = Q_i V_i V_{i+1} \cdots V_j | (Q_{j+1} L_{j+1})$$

where for every $i = 1, \ldots, m$, Q_i is a projection from L_i onto its first component.

Let $L = \sum_{i=1}^{m} L_i$, $Q = \sum_{i=1}^{m} Q_i$, all V_i be defined on L as partial isometries.

Let λ be a representation of $\tilde{M}$ on the space L defined as follows:

$$\lambda(I(a) \cdot E_{ij}) = \begin{cases} (\sigma \circ \rho)(I(a)) \cdot V_i V_{i+1} \cdots V_j & \text{if } i \leq j \\ (\sigma \circ \rho)(I(a)) \cdot (V_i V_{i+1} \cdots V_j)^* & \text{if } i > j \end{cases}$$

where $\sigma : \rho(I(M)) \to B(L)$ is the natural ampliation.

The map λ defined as above and extended by linearity is indeed a *-representation of $\tilde{M}$, and

$$\rho(T) = Q\lambda(T)|(QL), \qquad T \in \mathcal{T}$$

□

Corollary 18.11 *In the conditions of Lemma 18.10, $\mathcal{T}$ is universally completely contractive.*

PROOF. Let ρ be a contractive σ-weakly continuous representation of $\mathcal{T}$ on a Hilbert space K. Let $\tilde{M_1} = \tilde{M} \otimes \mathrm{Mat}_k$, $\mathcal{T}_1 = \mathcal{T} \otimes \mathrm{Mat}_k$, ρ_1 be the representation of $\mathcal{T}_1$ defined as follows:

$$\rho_1((a_{ij})) = (\rho(a_{ij})), \qquad a_{ij} \in \mathcal{T}$$

Let $M_1 = M \otimes \mathbb{C}^m$, where the multiplication in $\mathbb{C}^m$ is coordinatewise. Then $\tilde{M_1}$ is in effect the crossed product of M_1 by $R_1 = \{1, \ldots, mk\}^2$. The algebra $\mathcal{T}_1$ is a σ-Dirichlet subalgebra of $\tilde{M}_1$.

The map ρ_1 possesses the following property: for every matrix unit $E_{ij} \in \mathrm{Mat}_k$, and for every $T \in \mathcal{T}$: $\|\rho(T \otimes E_{ij})\| \leq \|T\|$. By the proof of Lemma 18.10, ρ_1 has a dilation. Hence ρ_1 is contractive.

□

Lemma 18.12 *Let, in the notation above, $X = S \times \{1, \ldots, m\}$, $R = \triangle_S \times \{1, \ldots, m\}^2$. Let $\mathcal{T} = \mathcal{T}(B)$ with*

$$B(x, y) = \begin{cases} 1_{H(y)} & \textit{if } (l, n) \in T \\ 0 & \textit{if } (l, n) \notin T \end{cases}$$

where $(x, y) \in R$, $x = (s, l)$, $y = (s, n)$, T is a fixed subset of $\{1, \ldots, m\}^2$.

Assume in addition that $\mathcal{T}$ is a σ-Dirichlet subalgebra of $\tilde{M}$ containing $I(M)$. Then $\mathcal{T}$ is universally completely contractive.

PROOF. Let $P_S = \chi_{\triangle_S \times \{1\}}$. Then $P_S \in Z(M)$. Let $Z_S = Z(M)_{P_S}$, $M_S = M_{P_S}$. Then $Z(M) \cong Z_S \otimes \mathbb{C}^m$, $M \cong M_S \otimes \mathbb{C}^m$, $\tilde{M} \cong I(M_S) \otimes \mathrm{Mat}_m$.

The algebra $\tilde{M}$ is in effect the crossed product of $M \cong \sum_{i=1}^{m} M_S$ by $\{1, \ldots, m\}^2$. By Corollary 18.11, $\mathcal{T}$ is universally completely contractive.

□

Lemma 18.13 *Let, in the notation above,* $X = S \times \{1, \ldots, m\}$, $R = \triangle_S \times \{1, \ldots, m\}^2$, $B \in \tilde{Z}^p$.

There exist disjoint projections $P_k \in Z_S$, $k = 1, \ldots, 2^{m^2}$, *such that* $\sum_{k=1}^{2^{m^2}} = P_S$, *and there exist subsets* $T_k \subseteq \{1, \ldots, m\}^2$, $k = 1, \ldots, 2^{m^2}$, *such that for a. e.* $z \in S$*:*

$$B[x, y] = \sum_{k:(l,n) \in T_k} \alpha_{((z,n),(z,1))}(P_k(z, 1))$$

where $x = (z, l)$, $y = (z, n)$, $z \in S$, $l, n \in \{1, \ldots, m\}^2$.

(Here P_S, Z_S *etc. are as in the proof of Lemma 18.12.)*

PROOF. For $\omega \subseteq \{1, \ldots, m\}^2$, $z \in S$ let:

$$\begin{aligned} P_\omega(z, 1) &= \prod \Big(\big\{ \alpha_{((z,1),(z,n))}(B[(z, l), (z, n)]) \mid (l, n) \in \omega \big\} \\ &\qquad \bigcup \big\{ P_S - \alpha_{((z,1),(z,n))}(B[(z, l), (z, n)]) \mid (l, n) \notin \omega \big\} \Big) \end{aligned}$$

Then:

1. $P_\omega \in Z_S$;
2. for $\omega_1 \neq \omega_2$: $P_{\omega_1} \cdot P_{\omega_2} = 0$;
3. $\sum_{\omega \subseteq \{1, \ldots, m\}^2} P_\omega = P_S$.

Indeed,

$$\begin{aligned} &\textstyle\sum_\omega P_\omega = \sum_\omega \prod(\{P(l, n) \mid (l, n) \in \omega\} \bigcup \{P_S - P(l, n) \mid (l, n) \notin \omega\}) \\ &= \textstyle\prod_{(l,n)} (P(l, n) + (P_s - P(l, n))) = \prod_{(l,n)} P_S = P_S \end{aligned}$$

where $P(l, n) \in Z_S$, $P(l, n) = \alpha_{((z,1),(z,n))}(B[(z, l), (z, n)])$. Furthermore,

$$B[(z, l), (z, n)] = \sum_{\omega:(l,n) \in \omega} P_\omega(z)$$

Indeed, for every $\omega \subseteq \{1, \ldots, m\}^2$ containing (l, n): $P_\omega(z) \leq B[(z, l), (z, n)]$, hence

$$\sum_{\omega:(l,n) \in \omega} P_\omega(z) \leq B[(z, l), (z, n)]$$

Furthermore, for every $\omega \subseteq \{1, \ldots, m\}^2$ not containing (l, n): $P_\omega(z) \leq P_S(z) - B[(z, l), (z, n)]$, hence

$$\sum_{\omega:(l,n) \notin \omega} P_\omega(z) \leq P_\omega(z) - B[(z, l), (z, n)]$$

The sum of two last inequalities is an equality. Hence, every of them is an equality. □

Lemma 18.14 *Let, in the notation above,* $X = S \times \{1, \ldots, m\}$, $R = \triangle_S \times \{1, \ldots, m\}^2$, $\mathcal{T}$ *be a* σ*-Dirichlet subalgebra of* $\tilde{M}$, *containing* $I(M)$. *Then* $\mathcal{T}$ *is universally completely contractive.*

PROOF. By the Spectral Theorem for Bimodules (Theorem 15.18), $\mathcal{T} = \mathcal{T}(B)$ for some $B \in \tilde{Z}_p$. By Lemma 18.13, there exist disjoint projections $P_k \in Z_S$, $k = 1, \ldots, 2^{m^2}$, and subsets $T_k \subseteq \{1, \ldots, m\}^2$, $k = 1, \ldots, 2^{m^2}$, such that:

$$B[(z,l),(z,n)] = \sum_{k:(l,n)\in T_k} \alpha_{((z,n),(z,1))}(P_k(z,1))$$

almost everywhere.

Let ρ be a σ-weakly continuous contractive representation of $\mathcal{T}$. For every $k = 1, \ldots, 2^{m^2}$, let ρ_k be defined as follows:

$$\rho_k(T) = \rho(I(\bar{P}_k)TI(\bar{P}_k)), \qquad T \in \mathcal{T}$$

where $\bar{P}_k \in Z(M)$, and for $z \in S$, $i \in \{1, \ldots, m\}$:

$$\bar{P}_k(s,i) = \alpha_{((s,i),(s,1))}(P_k(s,1))$$

Every ρ_k is in effect a representation of an algebra $\mathcal{T}_k = I(\bar{P}_k)\mathcal{T}I(\bar{P}_k)$, which is a subalgebra of an algebra

$$\tilde{M}_k = I(\bar{P}_k)\tilde{M}I(\bar{P}_k) \cong \tilde{M}_{I(\bar{P}_k)}$$

which in turn is the crossed product of $M_k = \bar{P}_k M \bar{P}_k \cong M_{\bar{P}_k}$ by R.

Furthermore, $\mathcal{T}_k$ is a σ-Dirichlet subalgebra of $\tilde{M}_k$, and $\mathcal{T}_k = \mathcal{T}(B_k)$ where $B_k \in \tilde{Z}^p$,

$$\begin{aligned} B_k[(z,l),(z,n)] &= \alpha_{((z,n),(z,1))}(P_k(z,1)) \cdot B[(z,l),(z,n)] \\ &= \begin{cases} \alpha_{((z,n),(z,1))}(P_k(z,1)) & \text{if } (l,n) \in T_k \\ 0 & \text{if } (l,n) \notin T_k \end{cases} = \begin{cases} 1_{H_k(z,n)} & \text{if } (l,n) \in T_k \\ 0 & \text{if } (l,n) \notin T_k \end{cases} \end{aligned}$$

By Lemma 18.12, every ρ_k is universally completely contractive.

Finally, by Lemma 18.9, ρ is universally completely contractive.

□

Lemma 18.15 *Let, in the notation above, every coset of* R *be finite,* $\mathcal{T}$ *be a* σ*-Dirichlet subalgebra of* $\tilde{M}$, *containing* $I(M)$. *Then* $\mathcal{T}$ *is universally completely contractive.*

PROOF. Let for $m \in \mathbb{N}$: $X_m = \{x \in X \mid \text{card}([x]) = m\}$. By [14, Theorem 2(a)], all X_m are Borel. Let $R_m = R \cap (X_m \times X_m)$. Then every R_m is a Borel equivalence relation, and $R = \bigcup_{m=1}^{\infty} R_m$.

Let $P_m = I(\chi_{X_m})$. Then $P_m \in Z(\tilde{M})$, all P_m are orthogonal projections, and $\sum_{m=1}^{\infty} P_m = 1_H$.

Let ρ be a σ-weakly continuous contractive representation of $\mathcal{T}$. For every $m \in \mathbb{N}$ let ρ_m be defined as follows:

$$\rho_m(T) = \rho(P_m T P_m), \qquad T \in \mathcal{T}$$

Every ρ_k is in effect a representation of an algebra $\mathcal{T}_m = P_m \mathcal{T} P_m$, which is a σ-Dirichlet subalgebra of the algebra

$$\tilde{M}_m = P_m \tilde{M} P_m \cong \tilde{M}_{P_m} = M_{\chi_{X_m}} \bowtie R_m$$

By Lemma 15.14 (or by [26, Lemma 4(1)]), $R_m \cong \triangle_S \times \{1, \dots, m\}^2$ for some Borel $S \subseteq X$, and $X \cong \{1, \dots, m\}$.

By Lemma 18.14, every ρ_m is universally completely contractive.

Finally, by [26, Lemma 2], ρ is universally completely contractive.

□

Theorem 18.16 (Theorem on Dilations) (See [26, Theorem 1].) *Let R be hyperfinite and $c \equiv 1$. Let $\mathcal{T}$ be a σ-Dirichlet subalgebra of $\tilde{M}$ containing $I(M)$. Then the pair $(\tilde{M}, \mathcal{T})$ is dilable, i. e. every σ-weakly continuous contraction ρ of $\mathcal{T}$ has a dilation to a normal *-representation of $\tilde{M}$.*

PROOF. It is enough to prove that every σ-weakly continuous contractive representation ρ of $\mathcal{T}$ is universally completely contractive.

Let $R = \bigcup_{n=1}^{\infty} R_n$ where every R_n is an equivalence relation with finite cosets, and $R_n \subset R_{n+1}$. For every $n \in \mathbb{N}$, $\chi_{R_n} \in \tilde{Z}$. It is clear that $\chi_{R_n} \circ \chi_{R_n} = \chi_{R_n} = \theta(\chi_{R_n})$ and $\chi_{R_n}[x, x] = 1_{H(x)}$ for $x \in X$. By Theorem 14.4, $\mathcal{T}(\chi_{R_n})$ is a von Neumann subalgebra of $\tilde{M}$, and there exists a conditional expectation Φ_n from $\tilde{M}$ onto $\mathcal{T}(\chi_{R_n})$ which is given by the formula

$$\Phi_n(T)(x, y) = \chi_{R_n}[x, y] T(x, y)$$

By part (6) of the proof of Theorem 15.18, $\Phi_n(T) \to T$ σ-strongly. By [35, Theorem 1], Φ_n are universally completely contractive. Furthermore, $\Phi_n(\mathcal{T}) \subseteq \mathcal{T}(\chi_{R_n})$.

Let $(a_{ij}) \in \mathcal{T} \otimes \mathrm{Mat}_k$. Then

$$\|(\rho(a_{ij}))\| \leq \limsup_n \|(\rho \circ \Phi_n(a_{ij}))\|$$

because $(\Phi_n(a_{ij})) \overset{\sigma w}{\to} (a_{ij})$, and ρ is σ-weakly continuous.

Every $\Phi_n(\mathcal{T})$ is a σ-Dirichlet subalgebra of $\mathcal{T}(\chi_{R_n})$. By Lemma 18.15, every $\Phi_n(\mathcal{T})$ is universally completely contractive, hence every $\rho \circ \Phi_n$ is universally completely contractive. Thus,

$$\|(\rho \circ \Phi_n(a_{ij}))\| \leq \|(a_{ij})\|$$

Hence, $\|(\rho(a_{ij}))\| \leq \|(a_{ij})\|$, i. e. ρ is universally completely contractive.

□

CHAPTER 19

THE CONSTRUCTION OF YAMANOUCHI

Crossed products of von Neumann algebras by groupoids were defined and studied by Takehiko Yamanouchi [37]. The "intersection" of our construction with the construction of Yamanouchi is the particular case of our construction, in which $c \equiv 1$, and the particular case of the construction of Yamanouchi, where the groupoid is an equivalence relation with countable cosets.

We shell show now that in this case our construction is essentially isomorphic to Yamanouchi's. The precise statement is Theorem 19.8 below.

Let us recall the construction of the crossed product of a von Neumann algebra by a groupoid action of Yamanouchi.

Let $\mathcal{G}$ be a measure groupoid. (See [17, Definitions 1.1 and 2.3].) Let X be its unit space. For every $y \in X$ let $\mathcal{G}^y = r^{-1}(y)$.

Definition 19.1 (See [8, Ch. I, Definition 2].) *A* **transverse function** *on $\mathcal{G}$ is the family $\{\lambda^u\}_{u\in X}$ of measures on $\mathcal{G}$ such that:*

1. *for every $u \in X$, the measure λ^u is supported on $\mathcal{G}^u$;*
2. *for every measurable $B \subset \mathcal{G}$, the function $u \mapsto \lambda^u(B)$ is measurable;*
3. *for every measurable positive function f on $\mathcal{G}$, a. e. $x \in \mathcal{G}$:*

$$\int_{\mathcal{G}} f(y)d\lambda^{r(x)}(y) = \int_{\mathcal{G}} f(xy)d\lambda^{d(x)}(y)$$

NOTATION. Let $\{\lambda^y\}_{y\in X}$ be a transverse function on $\mathcal{G}^y$.

Let μ be a quasiinvariant measure on X, $\nu = \int_X \lambda^y d\mu(y)$ be the corresponding measure on $\mathcal{G}$.

The groupoid $\mathcal{G}$ can be viewed as a category with inverses, so that $\mathrm{Obj}\,\mathcal{G} = X$, $\mathrm{Mor}\,\mathcal{G} = \mathcal{G}$, and every $\gamma \in \mathcal{G}$ is an arrow from $s(\gamma)$ to $r(\gamma)$.

Definition 19.2 (See [37, Definition 3.1].) *An* **action** *of $\mathcal{G}$ is a functor $\mathcal{F}$ from $\mathcal{G}$ to the category of von Neumann algebras with *-isomorphisms which is measurable in the following sense:*

1. *the family $\{\mathcal{F}(x), L^2(\mathcal{F}(x))\}_{x\in X}$ is a measurable field over (X, μ);*
2. *for every $a = \int a(x)d\mu(x) \in M = \int \mathcal{F}(x)d\mu(x)$ and for every $\omega = \int \omega_x d\mu(x) \in M_*$, the function $\gamma \mapsto \omega_{r(\gamma)}(\mathcal{F}(\gamma)(a(s(\gamma))))$ is ν-measurable $(\gamma \in \mathcal{G})$.*

Here for a von Neumann algebra A, the expression $L^2(A)$ means the canonical L^2-space associated with A by [21].

Let $\mathcal{F}$ be some acton of $\mathcal{G}$. Let $\bar{M}(x) = \mathcal{F}(x)$ for $x \in X$ and $\alpha_\gamma = \mathcal{F}(\gamma)$ for $\gamma \in \mathcal{G}$.

NOTATION. For every $x \in X$, let $\bar{H}(x)$ be the canonical Hilbert space associated with $\bar{M}(x)$. For $\gamma \in \mathcal{G}$, let $u(\gamma)$ be the canonical implementation of α_γ, so that $\alpha_\gamma = \operatorname{Ad} u(\gamma)$. Let

$$\begin{aligned} \hat{H}(x) &= \bar{H}(x) \otimes L^2(\mathcal{G}^x, \lambda^x), & x \in X \\ \hat{M}(\gamma) &= \{a \cdot u(\gamma) \otimes \lambda(\gamma) \mid a \in \bar{M}(r(\gamma))\}, & \gamma \in \mathcal{G} \end{aligned}$$

Here $\lambda(\gamma) : L^2(\mathcal{G}^x, \lambda^x) \to L^2(\mathcal{G}^x, \lambda^x)$ for $\gamma \in \mathcal{G}$, $\gamma : x \to y$ is defined as follows:

$$(\lambda(\gamma)\xi)(\gamma') = \xi(\gamma^{-1}\gamma'), \quad \xi \in L^2(\mathcal{G}^x, \lambda^x), \quad \gamma' \in \mathcal{G}$$

Let us consider the direct integral:

$$\hat{H} = \int_X \hat{H}(y) d\mu(y) = \int_X \bar{H}(y) \otimes L^2(\mathcal{G}^x, \lambda^x) d\mu(y)$$

This space is identified with the set of functions η from $\mathcal{G}$ into $\bigcup_{x \in X} \bar{H}(x)$, such that for $\gamma \in \mathcal{G}$: $\eta(\gamma) \in \bar{H}(r(\gamma))$ and satisfying a usual property of measurability and finiteness of L^2-norm. Thus, $\hat{H}$ is identified with the direct integral

$$\int_{\mathcal{G}} \bar{H}(\gamma) d\nu(\gamma)$$

where $\bar{H}(\gamma) \equiv \bar{H}(r(\gamma))$.

Let $S(\bar{M}) = S\left(\mathcal{G}, \bigcup_{\gamma \in \mathcal{G}} \hat{M}(\gamma)\right)$ be the set of sections A from $\mathcal{G}$ into $\bigcup_{\gamma \in \mathcal{G}}$, such that for $\gamma \in \mathcal{G}$: $A(\gamma) \in \hat{M}(\gamma)$ and satisfying the following properties:

1. the usual measurability: if A is of the form $A(\gamma) = a(\gamma) \otimes \lambda(\gamma)$, then for any $\omega = \int \omega_x d\mu(x) \in M_*$, the function $\gamma \mapsto \omega_{r(\gamma)}(a(\gamma))$ is measurable;
2. the quantity

$$\|A\|_H = \max\{\|\lambda(\|A(\cdot)\|)\|_\infty, \|\lambda(\|A^\#(\cdot)\|)\|_\infty\}$$

is finite.

Here for $\gamma \in \mathcal{G}$: $A^\#(\gamma) = \delta(\gamma)^{-1} A(\gamma^{-1})^*$, $\delta = d\nu/d\nu^{-1}$, and for a Borel function g on $\mathcal{G}$:

$$(\lambda(g))(x) = \int_{\mathcal{G}} g(\gamma) d\lambda^x(\gamma), \quad x \in X$$

Let us define the multiplication $*$ and the involution $\#$ on the algebra $S(\bar{M})$ as follows:

$$\begin{aligned} (A * B)(\gamma) &= \int_{\mathcal{G}} A(\gamma_1) B(\gamma_1^{-1}\gamma) d\lambda^{r(\gamma)}(\gamma_1), & A, B \in S(\bar{M}) \\ A^\#(\gamma) &= \delta(\gamma)^{-1} A(\gamma^{-1})^* \end{aligned}$$

Then $S(\bar{M})$ is a #-algebra.

Let us define the representation of $S(\bar{M})$ on the Hilbert space $\hat{H}$. For every $A \in S(\hat{M})$, $A(\gamma) = a(\gamma)u(\gamma) \otimes \lambda(\gamma)$ and for every $\xi, \eta \in \hat{H} \cong \int_{\mathcal{G}} \hat{H}(\gamma)d\nu(\gamma)$ let:

$$\langle \Phi(A)\xi, \eta\rangle = \int\int \langle a(\gamma_1)u(\gamma_1)\xi(\gamma_1^{-1}\gamma)\rangle d\lambda^{r(\gamma)}(\gamma_1)d\nu(\gamma)$$

Then $\Phi(A) \in B(\hat{H})$, and $\|\Phi(A)\| \leq \|A\|_H$, and Φ is non-degenerate contractive representation of $S(\bar{M}$ on $\hat{H}$. The following definition is the Yamanouchi's definition of the crossed product.

Definition 19.3 (See [37, the notation after Lemma 4.2].) *The weak closure of the image $\Phi(S(\bar{M}))$ of $S(\bar{M})$ is called the* **crossed product** *(in terms of Yamanouchi) of $\bar{M}$ by an action of $\mathcal{G}$ and denoted by*

$$\bar{M} \times_\alpha \mathcal{G}$$

(Here $\bar{M} = \int_X \bar{M}(x)d\mu(x)$.)

Now, let us return to the construction of Section 2. Let X, μ, M, H, R, ν, D, α etc. be as in Sections 2 and 3, and let $c(x, y, z) \equiv 1$.

We can assume that every coset of R is countable (not finite).

NOTATION. Let $\mathcal{G}_R = R$. Let $(x, y), (z, v) \in R$ be compatible (i. e. can be multiplied) if and only if $x = v$, and define:

$$(x, y) \cdot (z, x) = (z, y)$$

Let in addition $(x, y)^{-1} = (y, x)$.

Then $\mathcal{G}_R$ becomes a groupoid with the unit space $X_R = X$. All the notions concerning to this groupoid will be denoted by the subscript or superscript R.

From now on, for every function f defined on the groupoid $\mathcal{G}_R$ we will write $f(x, y)$ in place of $f((x, y))$.

We have:

$$s_R(x, y) = (x, y)^{-1} \cdot (x, y) = (y, x) \cdot (x, y) = (x, x) \cong x$$
$$r_R(x, y) = (x, y) \cdot (x, y)^{-1} = (x, y) \cdot (y, x) = (y, y) \cong y$$
$$\mathcal{G}^y = r_R^{-1}(y) = \{(z, y) \mid (z, y) \in R\}$$

Let λ_R^y be the counting measure on $\mathcal{G}^y$; it is invariant. Let $\mu_R = \mu$. Then

$$\nu_R = \int_X \lambda_R^y d\mu_R(y) = \int_X \operatorname{card}((\cdot) \cap r_R^{-1}(y))d\mu(y) = \nu;$$
$$\delta_R = d\nu_R/d\nu_R^{-1} = d\nu/d\nu^{-1} = D^{-1}$$

Let us define the action $\mathcal{F}_R$ of the groupoid $\mathcal{G}_R$ as follows:

$$\bar{M}(x) = \mathcal{F}_R(x) \stackrel{\text{def}}{=} M'(x) \quad \text{for} \quad x \in X = X_R;$$
$$\alpha^R_{(x,y)} = \mathcal{F}_R(x, y) \stackrel{\text{def}}{=} \alpha_{(y,x)} \quad \text{for} \quad (x, y) \in R = \mathcal{G}_R$$

Then $(x,y)\cdot(z,x) = (z,y)$, and $\alpha_{(y,x)}\circ\alpha_{(x,z)} = \alpha_{(y,z)}$. Thus, $\mathcal{F}_R$ is indeed a morphism.

Now, in place of the canonical Hilbert space $L^2(\mathcal{F}_R(x))$ we take the space $H(x)$ as above, with the cyclic and separating vector $\tilde{\phi}_0$.

Let $u_R(x,y) = U_{(y,x)}$. Then really

$$\alpha^R_{(x,y)} = \alpha_{(y,x)} = \operatorname{Ad} U_{(y,x)} = \operatorname{Ad} u_R(x,y)$$

In this notations, for every $y \in X$:

$$\begin{array}{rcl} L^2(\mathcal{G}^y_R, \lambda^y_R) & \cong & l^2(\mathbb{Z}) \\ \hat{H}_R(y) & = & H(y)\otimes l^2(\mathbb{Z}) \end{array}$$

Let us denote as above $H(x,y) \equiv H(y)$. Then $\hat{H}_R(y) = \sum_x H(x,y)$, and

$$\begin{array}{l} \hat{H}_R = \int_X \hat{H}_R(y)d\mu_R(y) = \int_X \sum_x H(x,y)d\mu(y) \\ = \int_R H(x,y)d\nu(x,y) = \tilde{H} \end{array}$$

We can now find the function λ_R.

$$(\lambda_R(\gamma_R)\xi)(\gamma'_R) = \xi(\gamma_R^{-1}\gamma'_R)$$

where $\gamma'_R \in \mathcal{G}_R$, $\gamma_R : x \mapsto y$.

Let $\gamma_R = (x,y)$, $\gamma'_R = (z,y)$. Then $\gamma_R^{-1}\gamma'_R = (y,x)\cdot(z,y) = (z,x)$. Thus,

$$(\lambda_R(x,y)\xi)(z,y) = \xi(z,x)$$

where $\xi \in \sum_z \mathbb{C}(z,x)$, $\mathbb{C}(z,x) \cong \mathbb{C}$.

So, the elements of $\hat{M}_R(x,y)$ map $\sum_z H(z,x)$ into $\sum_z H(z,y)$ by the rule:

$$\left(\left(aU_{(y,x)}\otimes\lambda_R(x,y)\right)\xi\right)(z,y) = aU_{(y,x)}\xi(z,x), \qquad a \in M'(y)$$

Now, let A be a section from $\mathcal{G}_R$ to $\bigcup_{(x,y)\in R}\hat{M}_R(x,y)$ such that for $(x,y)\in R$: $A(x,y)\in\hat{M}_R(x,y)$. We have:

$$\begin{array}{l} \|A\|_H = \max\left\{\|\lambda_R(\|A(\cdot)\|)\|_\infty, \left\|\lambda_R\left(\|A^{\#}(\cdot)\|\right)\right\|_\infty\right\} \\ = \max\left\{\sup_{u\in X}|\lambda_R(\|A(\cdot)\|)(u)|, \sup_{u\in X}\left|\lambda_R\left(\|A^{\#}(\cdot)\|\right)(u)\right|\right\} \\ = \max\left\{\sup_{u\in X}\left|\sum_v\|A(v,u)\|\right|, \sup_{u\in X}\left|\sum_v\|A^{\#}(v,u)\|\right|\right\} \\ = \max\left\{\sup_{u\in X}\sum_v\|A(v,u)\|, \sup_{u\in X}\sum_v\|A^{\#}(v,u)\|\right\} \end{array}$$

For every $T' \in \tilde{M}'$, let $T'(x,y)$ be the coordinates of T' as in Section 7. Let us denote

$$A_{T'}(x,y) = T'(x,y)U_{(y,x)}\otimes\lambda_R(x,y)$$

Thus for $\xi \in \sum_z H(z,x)$ we have:

$$(A_{T'}(x,y)\xi)(z,y) = T'(x,y)U_{(y,x)}\xi(z,x)$$

Then for $(x,y)\in R$: $\|A_{T'}(x,y)\| = \|T'(x,y)\|$.

For every $\tau\in\Sigma$, $\sigma\geq 1$ let τ_σ be a partial Borel homomorphism of X, defined as follows:

$$\mathcal{D}(\tau_\sigma) = \{x\in X \mid \frac{1}{\sigma}\leq D(x,\tau x)\leq\sigma\}$$

and for $x\in\mathcal{D}(\tau_\sigma)$, let $\tau_\sigma x = \tau x$.

Lemma 19.4 *For every* $a' \in M'$, $\tau \in \Sigma$, $\sigma \geq 1$: $A_{I'(a')}, A_{\pi'(\tau_\sigma)} \in S(\hat{M})_R$.

PROOF. The proof is straightforward.

□

Lemma 19.5 *Let* $A \in S(\bar{M})_R$, $A(\gamma) = a(\gamma)u_R(\gamma) \otimes \lambda_R(\gamma)$, $\gamma \in \mathcal{G}$. *Then for* $\xi \in \hat{H}_R \cong \tilde{H}$:

$$(\Phi_R(A)\xi)(x,y) = \sum_z a(z,y)U_{(y,z)}\xi(x,z)$$

PROOF. By the definition of $\Phi_R(A)$, for $\xi, \eta \in \hat{H}_R \cong \tilde{H}$:

$$\langle \Phi_R(A)\xi, \eta\rangle = \int\int \langle a(\gamma_1)u(\gamma_1)\xi(\gamma_1^{-1}\gamma), \eta(\gamma)\rangle d\lambda^{r(\gamma)}(\gamma_1)d\nu(\gamma)$$

Let $\gamma = (x,y)$, $\gamma_1 = (z,y)$. Then $\gamma_1^{-1}\gamma = (y,z) \cdot (x,y) = (x,z)$, and we have:

$$\begin{aligned}\langle \Phi_R(A)\xi, \eta\rangle &= \textstyle\int_R \sum_z \langle a(z,y)U_{(y,z)}\xi(x,z), \eta(x,y)\rangle d\nu(x,y)\\ &= \textstyle\int_R \langle \sum_z a(z,y)U_{(y,z)}\xi(x,z), \eta(x,y)\rangle d\nu(x,y)\end{aligned}$$

(by the Lebeque convergence theorem). Thus,

$$(\Phi_R(A)\xi)(x,y) = \sum_z a(z,y)U_{(y,z)}\xi(x,z)$$

□

Corollary 19.6 *If for some* $T' \in \tilde{M}'$: $A_{T'} \in S(\bar{M})_R$, *then* $\Phi_R(A_{T'}) = T'$.

PROOF. For $\xi \in \hat{H}_R \cong \tilde{H}$:

$$(\Phi_R(A_{T'})\xi)(x,y) = \sum_z T'(z,y)U_{(y,z)}\xi(x,z) = (T'\xi)(x,y)$$

□

Corollary 19.7 *For* $a' \in M'$, $\tau \in \Sigma$, $\sigma \geq 1$:

$$\begin{aligned}\Phi_R(A_{I'(a')}) &= I'(a')\\ \Phi_R(A_{\pi'(\tau_\sigma)}) &= \pi'(\tau_\sigma)\end{aligned}$$

The relationship between our definition of a crossed product, $\tilde{M}$, and Yamanouchi's [37] is shown in the following theorem.

Theorem 19.8 $M' \times_\alpha \mathcal{G}_R = \tilde{M}'$

PROOF. "$\supset$". The algebra $\tilde{M}$ is generated by all of $I'(a')$ with $a' \in M'$ and all of $\pi'(\tau_\sigma)$ with $\tau \in \Sigma$, $\sigma \geq 1$.

By the definition and by Corollary 19.6, the algebra $M' \times_\alpha \mathcal{G}_R$ contains all of $I'(a')$ with $a' \in M'$ and all of $\pi'(\tau_\sigma)$ with $\tau \in \Sigma$, $\sigma \geq 1$. Hence, $M' \times_\alpha \mathcal{G}_R$ contains $\tilde{M}$.

"$\subset$". To prove the converse inclusion it is enough to check that for every $A \in S(\bar{M})_R$, the operator $\Phi_R(A)$ commutes with elements of $\tilde{M}$. For this, in turn, it is enough to check that $\Phi_R(A)$ commutes with $I(b)$ for $b \in M$ and with $\pi(\tau)$ for $\tau \in \Sigma$.

Let $A \in S(\bar{M})_R$, $A(\gamma) = a'(\gamma)u_R(\gamma) \otimes \lambda_R(\gamma)$, $\gamma \in \mathcal{G}_R$. Then by Lemma 19.5:

$$(\Phi_R(A)\xi)(x, y) = \sum_z a'(z, y)U_{(y,z)}\xi(x, z)$$

where $a'(z, y) \in M'(y)$. The commutativity is verified straightforward.

$\square$

CHAPTER 20

EXAMPLES AND PARTICULAR CASES

20.1 The crossed product of a von Neumann algebra by an A-free group of automorphisms

Let M be a von Neumann algebra with a separable predual, acting on a Hilbert space H and possessing a cyclic and separating vector ϕ_0. Let G be a discrete countable group with unit e. Let α be an **action** of G, i. e. $\alpha_{(\cdot)} : G \to \text{Aut}M$ be a homomorphism of G into the group of $*$-automorphisms of the algebra M.

Definition 20.1 (See [19, Definition 13.1.1].) *The* **crossed product** *of the von Neumann algebra M by the action α of the group G is the von Neumann algebra acting on the Hilbert space $H \otimes l^2(G)$ generated by operators*

$$\sum_{g\in G}\Big(\alpha_g^{-1}(a)\otimes E_g\Big), \qquad a \in M$$

$$1_H \otimes l_g, \qquad g \in G$$

where for $g \in G$: $E_g, l_g \in B(l^2(G))$, and for $\xi \in l^2(G)$:

$$\begin{aligned}(E_g\xi)(h) &= \delta_{g,h}\xi(h)\\ (l_g\xi)(h) &= \xi(g^{-1}h)\end{aligned}$$

The crossed product of a von Neumann algebra M by an action α of a group G is denoted by

$$M \times_\alpha G$$

Let A be an Abelian von Neumann subalgebra of M (for example, A may be the center of M). Let $A \cong L^\infty(X,\mu)$ where (X,μ) is a standard Borel space. Let for every $g \in G$: $\alpha_g(A) = A$ (if $A = Z(M)$, then the last follows from $\alpha_g(M) = M$).

Let $M = \int_X M(x)d\mu(x)$ and $H = \int_X H(x)d\mu(x)$ be the decompositions of M and H into direct integrals corresponding to the representation $A \cong L^\infty(X,\mu)$.

Definition 20.2 *The action α of the group G will be called* **A-free**, *if an equation $\alpha_g(a) = a$ for some $a \in A$, $g \in G$, $g \neq e$ implies $a = c\cdot 1_H$, $c \in \mathbb{C}$.*

Let the action α be A-free.

Without loss of generality we can assume that for every $g \in G$, the automorphism α_g is implemented by a unitary operator U_g, such that $g \mapsto U_g$ is the unitary representation of G.(See [19, Proposition 13.1.2].)

Let for every $g \in G$, the automorphism α_g of the Abelian algegra A corresponds to a Borel automorphism τ_g of the space (X, μ), i. e. for $a \in A$, a. e. $x \in X$:

$$(\alpha_g(a))(x) = a(\tau_g^{-1}x)$$

REMARK. For every $g \in G$, $g \neq e$, a. e. $x \in X$: $\tau_g x \neq x$. Indeed, assume that there exist $g \in G$, $E \subset X$, $\mu(E) > 0$ such that for every $x \in E$: $\tau_g x = x$. Then $\alpha_g(\chi_E) = \chi_E$ — this is a contradiction to the "A-freedom".

By [16, Proposition 4.1], there exist unitary operators $U_{g,x} : H(x) \to H(\tau_g x)$ such that for $\xi \in H$, a. e. $x \in X$:

$$(U_g\xi)(x) = r(\tau_g^{-1})^{1/2} U_{g,\tau_g^{-1}x} \xi(\tau_g^{-1}x)$$

where

$$r(x) = \frac{d\mu(\tau_g x)}{d\mu(x)}$$

The straight calculation shows that for $a \in M$, $f \in H$, a. e. $x \in X$:

$$\begin{aligned}(\alpha_g(a)f)(x) &= (U_g a U_g^* f)(x) \\ &= U_{g,\tau_g^{-1}x} a(\tau_g^{-1}x) U_{g,\tau_g^{-1}x} * f(\tau_g^{-1}x)\end{aligned}$$

Let us denote $\alpha_{g,\tau_g^{-1}x} = \operatorname{Ad} U_{g,\tau_g^{-1}x}$. Then

$$\begin{aligned}&(\alpha_g(a))(x) = \alpha_{g,\tau_g^{-1}x}(a(\tau_g^{-1}x)); \\ &(\alpha_g^{-1}(a))(y) = (\alpha_{g^{-1}}(a))(y) = \alpha_{g^{-1},\tau_g x}(a(\tau_g y))\end{aligned}$$

Let

$$R = \{(x,y) \mid x, y \in X, \text{ there exists } g \in G : \tau_g y = x\}$$

For $(x,y), (y,z) \in R$ let:

$$\begin{aligned} c(x,y,z) &\equiv 1_{H(z)}; \\ \alpha_{(y,x)} &= \alpha_{g^{-1},\tau_g y} \end{aligned}$$

where $g \in G$ is such that $\tau_g y = x$ (by the Remark above, there exists only one such g). Then $\alpha_{(y,x)}$ acts from $M(x)$ onto $M(y)$.

Lemma 20.3 *For a. e.* $x \overset{R}{\sim} y \overset{R}{\sim} z$:

$$\alpha_{(x,y)} \circ \alpha_{(y,z)} = \alpha_{(x,z)}$$

PROOF. Let $\tau_g y = x$, $\tau_h z = y$, $g, h \in G$. Then:

$$\begin{aligned} \alpha_{(z,y)} &= \alpha_{h^{-1},\tau_h z} \\ \alpha_{(y,x)} &= \alpha_{g^{-1},\tau_g z} \\ \alpha_{(z,x)} &= \alpha_{(gh)^{-1},\tau_{gh} z} \end{aligned}$$

because $x = \tau_g y = \tau_g \tau_h z = \tau_{gh} z$. It is easy to see that for every $a \in A$:

$$\alpha_{(gh)^{-1},\tau_{gh}z}(a(\tau_{gh}z))$$
$$= \alpha_{h^{-1},\tau_h z}(\alpha_{g^{-1},\tau_g y}(a(\tau_{gh}z)))$$

Thus,

$$\alpha_{(gh)^{-1},\tau_{gh}z} = \alpha_{h^{-1},\tau_h z} \circ \alpha_{g^{-1},\tau_g y}$$

□

Let $\tilde{M} = M \bowtie R$ in our sense.

Furthermore, let W act from the space $H \otimes l^2(G) \cong l^2(G, H)$ to our space $\tilde{H}$ as follows: for $f \in H \otimes l^2(G)$,

$$(Wf)(x, y) = f(g, y)$$

Then W is unitary. Furthermore, W satisfy the following properties:

$$\begin{aligned} W \sum_{g \in G} \left(\alpha_g^{-1}(a) \otimes E_g\right) W^* &= I(a); \\ W(1_H \otimes l_g)W^* &= \pi(\tau_g^{-1}) \end{aligned}$$

Thus, the operator W implements a spatial isomorphism of the algebra $M \times_\alpha G$ onto our algebra $\tilde{M} = M \bowtie R$.

REMARK. The condition of the A-free action of the group is essential here. Example: let $M = \mathbb{C}$, $H = \mathbb{C}$, $G = \{0, 1\}$ — the group of residues modulo 2, $\alpha_0 = \alpha_1 = \mathrm{id}_M$. Then we have:

$$M \times_\alpha G = \left\{ \begin{pmatrix} a & b \\ b & a \end{pmatrix} \;\middle|\; a, b \in \mathbb{C} \right\}$$

but this algebra can not be implemented as our crossed product of some von Neumann algebra by some equivalence relation.

20.2 Crossed product by a hyperfinite equivalence relation

Let as above $M = \int_X M(x) d\mu(x)$, R be a Borel hyperfinite equivalence relation on X.

By [14, Proposition 4.1], $R = \bigcup_{n=1}^{\infty} \Gamma(\tau^n)$ for some Borel automorphism τ of X.

Let $X_1 = \{x \in X \mid \tau x = x\}$. Let for $i = 2, 3, \ldots$: $Y_i = \{x \in X \mid \tau^i x = x\}$, $X_i = Y_i \setminus \bigcup_{j=1}^{i-1} X_j$, and finally $X_0 = X \setminus (\bigcup_{i=1}^{\infty} X_i)$. Let also $R_i = R \cap (X_i \times X_i)$ for $i = 0, 1, 2, \ldots$ Then every R_i is an equivalence relation on X_i. in particular: $R_1 = \{(x, x) \mid x \in X_1\}$.

Let for $i = 0, 1, 2, \ldots$: $M_i = \int_{X_i} M(x) d\mu(x)$. Then $M = \sum_{i=0}^{\infty} M_i$.

Let $\tilde{M} = M \bowtie R$ with some α and c as above. It is clear that $\tilde{M} = \sum_{i=0}^{\infty} \tilde{M}_i$, where

$$\tilde{M}_i = M_i \bowtie R_i, \qquad i = 0, 1, 2, \ldots$$

Let $\alpha_\tau \in \mathrm{Aut}(M)$ be defined as follows:

$$\alpha_\tau(a)(x) = \alpha_{(x,\tau^{-1}x)}(a(\tau^{-1}x))$$

Then $\theta : n \mapsto (\alpha_\tau)^n$ is the representation of $\mathbb{Z}$ as a group of automorphisms of $\tilde{M}$. Moreover, for $i = 0, 1, 2, \ldots$: $\alpha_\tau(M_i) = M_i$. Hence, for every $i = 0, 1, 2, \ldots$, the map $\theta_i : n \mapsto (\alpha_\tau)^n | \tilde{M}_i$ is the representation of $\mathbb{Z}$ as group of automorphisms of $\tilde{M}_i$.

Moreover, for every $i = 1, 2, \ldots$, the map θ_i is in fact the representation of the group $\mathbb{Z}_n$ of residues modulo n, and this representation is A-free (see Subsection 20.1). The map θ_0 in turn is the A-free representation of $\mathbb{Z}$.

By Subsection 20.1, for $i = 1, 2, \ldots$: $M_i \bowtie R_i \cong M_i \times_{\theta_i} \mathbb{Z}_i$. Thus, in our case

$$M \bowtie R \cong \sum_{i=1}^{\infty} (M_i \times_{\theta_i} \mathbb{Z}_i) \oplus (M_0 \times_{\theta_0} \mathbb{Z})$$

REMARK. It is clear that for $i = 1, 2, \ldots$: $R_i = \{x \in X \mid \mathrm{card}([x]) = i\}$, and $R_0 = \{x \in X \mid \mathrm{card}([x]) = \infty\}$. Thus, if (almost) every coset of R if countable (not finite), then $M = M_0$, $R = R_0$, and

$$M \bowtie R \cong M \times_{\theta_0} \mathbb{Z}$$

20.3 Double crossed product

Let as above $M = \int M(x) d\mu(x)$, $\tilde{M} = M \bowtie R$ is the crossed product built by a set of correspondence maps $\alpha_{(\cdot,\cdot)}$ and a "cocycle" $c(\cdot,\cdot,\cdot)$.

Let for a. e. $x \in X$:

$$M(x) = L_x \bowtie Q_x \tag{20.1}$$

where $L_x = \int_{S_x} L_x(s) d\lambda(s)$ is a decomposition of L_x into the direct integral, Q_x is an equivalence relation on S_x, and the crossed product (20.1) is built by a set of correspondence maps $\beta^x_{(\cdot,\cdot)}$ and a "cocycle" $d_x(\cdot,\cdot,\cdot)$. We will denote all the objects concerning this crossed product by subscript or superscript x.

Let $K_x = \int_{S_x} K_x(s) d\lambda_x(s)$ be the Hilbert space on which L_x acts.

Let κ_x be the measure on Q_x defined as usual by:

$$\kappa_x(B) = \int_{S_x} \mathrm{card}(B \cap r_x^{-1}(t)) d\lambda_x(t)$$

where for $(s,t) \in Q_x$: $r_x(s,t) = t$. Then

$$H(x) = \int_{Q_x} K_x(s,t) d\kappa_x(s,t)$$

where $K_x(s,t) \equiv K_x(t)$.

Let in addition:

1. for a. e. $(x,y) \in R$, the isomorphism $\alpha_{(y,x)} : M(x) \to M(y)$ is an I-isomorphism (see Subsection 13.2), corresponding to the Borel isomorphism $\tau_{(y,x)} : S_x \to S_y$;

2. for a. e. $x \sim y \sim z$, the operator $c(x,y,z)$ commutes with every $b \in B_x$.

We can assume that $S_x \cap S_y = \emptyset$ for a. e. $(x,y) \in R$.
Let us define:

1. $\bar{X} = \bigcup_{x \in X} S_x$;

2. $\bar{R} = \{(\bar{x},\bar{y}) \in \bar{X} \times \bar{X} \mid \bar{X} \overset{Q_x}{\sim} \tau^{-1}_{(y,x)}\bar{y}$ where $x, y \in X$ are such that $\bar{x} \in S_x$, $\bar{y} \in S_y\}$.

Then $\bar{R}$ is an equivalence relation on $\bar{X}$ with countable cosets.
Let $U_{(y,x)}$ be as above, i. e. such that $\alpha_{(y,x)} = \operatorname{Ad} U_{(y,x)}$. Let

$$r_{(y,x)}(q) = \frac{d(\kappa_y \circ \tau^{(2)}_{(y,x)})}{d\kappa_x}(q), \qquad q \in Q_x$$

Then by Lemmas 13.10 and 13.7, there exists unitary operators $U_{(y,x)}(q)$ from $K_x(q)$ onto $K_y(\tau^{(2)}_{(y,x)}q)$ such that for $\xi \in H(x)$, a. e. $p \in Q_y$:

$$(U_{(y,x)}\xi)(p) = r_{(y,x)}(\tau^{-1}_{(y,x)}p)\xi(\tau^{-1}_{(y,x)}p)$$

By the property of $c(x,y,z)$, the automorphism implemented by this operator corresponds to the identity isomorphism of Q_z. By [16, Proposition 4.1], for $\xi \in H(z)$:

$$(c(x,y,z)\xi)(q) = c(x,y,z)(q)\xi(q), \qquad q \in Q_z$$

where $c(x,y,z)(q)$ are unitary operators.
Now, let us define:

$$\begin{aligned} \bar{\alpha}_{(\bar{y},\bar{x})} &= \operatorname{Ad}\left(U_{(y,x)}(\bar{x}, \tau^{-1}_{(y,x)}\bar{y})\right) \circ \beta^{x}_{(\tau^{-1}_{(y,x)}\bar{y},\bar{x})}; \\ \bar{c}(\bar{x},\bar{y},\bar{z}) &= U_{(y,x)}(\bar{x}, \tau^{-1}_{(y,x)}\bar{y}) d_x(\bar{x}, \tau^{-1}_{(z,x)}\bar{z}, \tau_{(y,x)}\bar{y}) U_{(y,x)}(\bar{x}, \tau^{-1}_{(y,x)}\bar{y})^* \\ &\qquad c(x,z,y)(\tau_{(y,z)}\bar{z},\bar{y}) \end{aligned}$$

(Note: $L_x(\bar{x}) \overset{\beta_{(\tau^{-1}_{(y,x)}\bar{y},\bar{x})}}{\longrightarrow} L_x(\tau^{-1}_{(y,x)}\bar{y}) \overset{\alpha_{(y,x)}(\bar{y})}{\longrightarrow} L_y(\bar{y})$.)
Then $\bar{\alpha}$ is the set of correspondence maps as above, and $\bar{c}$ is the "cocycle" as above.
Let for $\bar{x} \in \bar{X}$: $\bar{M}(\bar{x}) \equiv M_x(\bar{x})$, where $x \in X$ is such that $\bar{x} \in S_x$. Let $\bar{M} = \int_{\bar{X}} \bar{M}(\bar{x}) d\bar{\mu}(\bar{x})$.
Let $\tilde{\bar{M}} = \bar{M} \bowtie \bar{R}$ by $\bar{\alpha}$ and $\bar{c}$.

Theorem 20.4 *Under the notation above, $\tilde{\bar{M}}$ is spatially isomorphic to $\tilde{M}$.*

PROOF. Let $\bar{H}(\bar{x}) = K_x(\bar{x})$, where $x \in X$ is such that $\bar{x} \in S_x$. Let $\bar{H}(\bar{x},\bar{y}) = \bar{H}(\bar{y})$. Let

$$\begin{aligned} \bar{H} &= \int_{\bar{X}} \bar{H}(\bar{x}) d\bar{\mu}(\bar{x}); \\ \bar{\bar{H}} &= \int_{\bar{R}} \bar{H}(\bar{x},\bar{y}) d\bar{\nu}(\bar{x},\bar{y}) \end{aligned}$$

where $\bar{\nu}$ is defined as above: $\bar{r}(\bar{x},\bar{y}) = \bar{y}$ and

$$\bar{\nu} = \int_{\bar{X}} \operatorname{card}\left((\cdot) \cap \bar{r}^{-1}(\bar{y})\right) d\bar{\mu}(\bar{y})$$

Then $\bar{M}$ acts on $\bar{H}$, and $\bar{\tilde{M}}$ acts on $\bar{\bar{H}}$.

Let $W : \tilde{H} \to \bar{\bar{H}}$ be defined as follows:

$$(W\xi)(\bar{x},\bar{y}) = \xi(x,y)(\tau_{(y,x)}\bar{x},\bar{y})$$

(here we assume that $\xi(x,y) \in H(x,y) \equiv H(y) = \int_{Q_y} K_y(s,t) d\kappa_y(s,t)$.) Then

$$(W^{-1}\xi)(x,y) = \int_{Q_y} \xi(\tau_{(y,x)}^{-1}s,t) d\kappa_y(s,t)$$

Moreover, the straight calculation shows that W is an isometry. Moreover, W acts from $\tilde{H}$ onto $\bar{\bar{H}}$, so W is a unitary operator.

To prove the theorem it is enough to show that $W\tilde{M}W* = \bar{\tilde{M}}$.

"$\subset$": Let $\xi \in \bar{\bar{H}}$, $(\bar{x},\bar{y}) \in \bar{R}$, $\bar{x} \in S_x$, $\bar{y} \in S_y$, $a \in M$, $\rho \in \Sigma$. It is enough to prove that $WI(a)\pi(\rho)W^* \in \bar{\tilde{M}}$. Furthermore, it is enough to take the case in which for a. e. $x \in X$: $a(x) = I_x(b_x)\pi_x(\sigma_x)$, where the indices x mean the usual objects concerning the algebra $M(x) = L_x \bowtie Q_x$.

The straight calculation shows that in this case:

$$\begin{aligned} &(WI(a)\pi(\rho)W^*\xi)(\bar{x},\bar{y}) \\ &= (\bar{I}(\bar{a})\bar{\pi}(\bar{\rho})\xi)(\bar{x},\bar{y}) \end{aligned}$$

where $\bar{a} \in \bar{M}$ and $\bar{\rho} \in \Sigma$ are defined by:

$$\bar{a}(\bar{x}) = b_x(\bar{x}) \tag{20.2}$$

$$\bar{\rho}\bar{x} = \tau_{(y,\rho x)}^{-1}\tau_{(y,x)}\sigma_x\bar{x} \tag{20.3}$$

"$\supset$": Let $\xi \in \bar{\bar{H}}$, $(\bar{x},\bar{y}) \in \bar{R}$, $\bar{x} \in S_x$, $\bar{y} \in S_y$, $\bar{a} \in \bar{M}$, $\bar{\rho} \in \bar{\Sigma}$. It is enough to prove that $W^*\bar{I}(\bar{a})\bar{\pi}(\bar{\rho})W \in \tilde{M}$.

We can choose $b_x \in L_x$, $\rho \in \Sigma$ and $\sigma_x \in \Sigma_x$ such that the equalities (20.2) and (20.3) will be satisfied. Then, as in the first part of this proof,

$$W^*\bar{I}(\bar{a})\bar{\pi}(\bar{\rho})W = I(a)\pi(\rho)$$

where as above $a(x) = I_x(b_x)\pi_x(\sigma_x)$.

□

BIBLIOGRAPHY

[1] W. Arveson. *Subalgebras of C^*-algebras.* Acta Math., **123** (1969), 141–224.

[2] W. Arveson. *On groups of automorphisms of operator algebras.* J. Func. Anal., **15** (1974), 217–243.

[3] W. Arveson. *Operator algebras and invariant subspaces.* Ann. of Math., **100** (1974), 433–532.

[4] L. Auslander and C. C. Moore. *Unitary representations of solvable Lie groups.* Mem. Amer. Math. Soc., **62** (1966), 199pp.

[5] N. Bourbaki. *Intégration*, chapitre VI. Hermann, Paris, 1959.

[6] O. Bratteli and D. W. Robinson. *Operator Algebras and Quantum Statistical Mechanics*, volume I. Springer-Verlag, 1979.

[7] A. Connes. *Classification of injective factors.* Ann. of Math., **104** (1976), 73–115.

[8] A. Connes. *Sur la théorie non commutative de l'intégration.* Lecture Notes in Math., **725** (1979), 19–143.

[9] A. Connes, J. Feldman, and B. Weiss. *An amenable equivalence relation is generated by a single transformation.* Ergod. Th. and Dynam. Sys., **1** (1981), 431–450.

[10] J. Dixmier. *Les Algèbres d'Opérateurs dans l'Espace Hilbertien (Algèbres de Von Neumann).* Gauthier-Villars, 1969.

[11] R. G. Douglas and V. I. Paulsen. *Hilbert modules over function algebras.* Pitman Res. Notes in Math., **217** (1989), 130 pp.

[12] G. A. Elliott. *On approximately finite-dimensional von Neumann algebras*, II. Canad. Math. Bull., **21** (1978), 415–418.

[13] G. A. Elliott and E. J. Woods. *The equivalence of various definitions for a property infinite von Neumann algebra to be approximately finite dimensional.* Proc. Amer. Math. Soc., **60** (1976), 175–178.

[14] J. Feldman and C. C. Moore. *Ergodic equivalence relations, cohomology, and von Neumann algebras*, I. Trans. Amer. Math. Soc., **234** (1977), 289–324.

[15] J. Feldman and C. C. Moore. *Ergodic equivalence relations, cohomology, and von Neumann algebras*, II. Trans. Amer. Math. Soc., **234** (1977), 325–359.

[16] A. Guichardet. *Une caractérisation des algèbres de von Neumann discrètes.* Bull. Soc. math. France, **89** (1961), 77–101.

[17] P. Hahn. *Haar measure for measure groupoids.* Trans. Amer. Math. Soc., **242** (1978), 1–33.

[18] E. Hewitt and K. A. Ross. *Abstract Harmonic Analysis*, volume I. Springer-Verlag, 1963.

[19] R. V. Kadison and J. R. Ringrose. *Fundamentals of the Theory of Operator Algebras*, volume II. Academic Press, 1986.

[20] J. L. Kelley. *General Topology.* Van Nostrand, Princeton, N. J., 1957.

[21] H. Kosaki. *Positive Cones and Canonical L^p-spaces Associated with an Arbitrary Abstract Von Neumann Algebra.* Dissertation, UCLA, 1980.

[22] R. I. Loebl and Paul S. Muhly. *Analyticity and flows in von Neumann algebras.* J. Func. Anal., **29** (1978), 214–252.

[23] G. W. Mackey. *Borel structure in groups and their duals.* Trans. Amer. Math. Soc., **85** (1957), 134–165.

[24] C. C. Moore. *Group extensions and cohomology for locally compact group*, III. Trans. Amer. Math. Soc., **221** (1976), 1–33.

[25] Paul S. Muhly, Kichi-Suke Saito, and Baruch Solel. *Coordinates for triangular operator algebras.* Ann. of Math., **127** (1988), 245–278.

[26] Paul S. Muhly and Baruch Solel. *Dilations for representations of triangular algebras.* Bull. London Math. Soc., **21** (1989), 489–495.

[27] Paul S. Muhly and Baruch Solel. *Subalgebras of groupoid C^*-algebras.* J. reine angew. Math., **402** (1989), 41–75.

[28] Paul S. Muhly and Baruch Solel. *Dilations and commutant lifting for subalgebras of groupoid C^*-algebras.* Intern. J. Math., **5** (1994), 87–123.

[29] F. J. Murray and J. Von Neumann. *On rings of operators*, IV. Ann. of Math. (2), **44** (1943), 716–808.

[30] V. I. Paulsen, S. C. Power, and J. D. Ward. *Semidiscreteness and dilation theory for nest algebras.* J. Func. Anal., **80** (1988), 76–87.

[31] J. Peters, Y. Poon, and B. Wagner. *Triangular AF algebras.* J. Operator Theory, **23** (1990), 81–114.

[32] Ş. Strătilă. *Modular Theory in Operator Algebras.* Editura Academiei and Abacus Press, 1981.

[33] Ş. Strătilă and L. Zsidó. *Lectures on Von Neumann Algebras.* Editura Academiei and Abacus Press, 1979.

[34] M. Takesaki. *Conditional expectations in von Neumann algebras.* J. Func. Anal., **9** (1972), 306–321.

[35] J. Tomiyama. *On the product projection of norm one in the direct product of operator algebras.* Tôhoku Math. J. (2), **11** (1959), 305–313.

[36] S. Yamagami. *On the ideal structure of C^*-algebras over locally compact groupoids.* Preprint, 31 pp.

[37] T. Yamanouchi. *Duality for actions and coactions of measured groupoids on von Neumann algebras.* Mem. Amer. Math. Soc., **484** (1993), 109 pp.

DEPARTMENT OF MATHEMATICS, UNIVERSITY OF IOWA, IOWA CITY, IA 52242, USA
E-mail address: `ifulman@math.uiowa.edu`

Editorial Information

To be published in the *Memoirs*, a paper must be correct, new, nontrivial, and significant. Further, it must be well written and of interest to a substantial number of mathematicians. Piecemeal results, such as an inconclusive step toward an unproved major theorem or a minor variation on a known result, are in general not acceptable for publication. *Transactions* Editors shall solicit and encourage publication of worthy papers. Papers appearing in *Memoirs* are generally longer than those appearing in *Transactions* with which it shares an editorial committee.

As of September 30, 1996, the backlog for this journal was approximately 7 volumes. This estimate is the result of dividing the number of manuscripts for this journal in the Providence office that have not yet gone to the printer on the above date by the average number of monographs per volume over the previous twelve months, reduced by the number of issues published in four months (the time necessary for preparing an issue for the printer). (There are 6 volumes per year, each containing at least 4 numbers.)

A Copyright Transfer Agreement is required before a paper will be published in this journal. By submitting a paper to this journal, authors certify that the manuscript has not been submitted to nor is it under consideration for publication by another journal, conference proceedings, or similar publication.

Information for Authors and Editors

Memoirs are printed by photo-offset from camera copy fully prepared by the author. This means that the finished book will look exactly like the copy submitted.

The paper must contain a *descriptive title* and an *abstract* that summarizes the article in language suitable for workers in the general field (algebra, analysis, etc.). The *descriptive title* should be short, but informative; useless or vague phrases such as "some remarks about" or "concerning" should be avoided. The *abstract* should be at least one complete sentence, and at most 300 words. Included with the footnotes to the paper, there should be the 1991 *Mathematics Subject Classification* representing the primary and secondary subjects of the article. This may be followed by a list of *key words and phrases* describing the subject matter of the article and taken from it. A list of the numbers may be found in the annual index of *Mathematical Reviews*, published with the December issue starting in 1990, as well as from the electronic service e-MATH [**telnet e-MATH.ams.org** (or **telnet 130.44.1.100**). Login and password are **e-math**]. For journal abbreviations used in bibliographies, see the list of serials in the latest *Mathematical Reviews* annual index. When the manuscript is submitted, authors should supply the editor with electronic addresses if available. These will be printed after the postal address at the end of each article.

Electronically prepared papers. The AMS encourages submission of electronically prepared papers in AMS-TEX or AMS-LATEX. The Society has prepared author packages for each AMS publication. Author packages include instructions for preparing electronic papers, the *AMS Author Handbook*, samples, and a style file that generates the particular design specifications of that publication series for both AMS-TEX and AMS-LATEX.

Authors with FTP access may retrieve an author package from the Society's Internet node `e-MATH.ams.org` (130.44.1.100). For those without FTP access, the author package can be obtained free of charge by sending e-mail to `pub@math.ams.org` (Internet) or from the Publication Division, American Mathematical Society, P.O. Box 6248, Providence, RI 02940-6248. When requesting an author package, please specify $\mathcal{A}\mathcal{M}\mathcal{S}$-TeX or $\mathcal{A}\mathcal{M}\mathcal{S}$-LaTeX, Macintosh or IBM (3.5) format, and the publication in which your paper will appear. Please be sure to include your complete mailing address.

Submission of electronic files. At the time of submission, the source file(s) should be sent to the Providence office (this includes any TeX source file, any graphics files, and the DVI or PostScript file).

Before sending the source file, be sure you have proofread your paper carefully. The files you send must be the EXACT files used to generate the proof copy that was accepted for publication. For all publications, authors are required to send a printed copy of their paper, which exactly matches the copy approved for publication, along with any graphics that will appear in the paper.

TeX files may be submitted by email, FTP, or on diskette. The DVI file(s) and PostScript files should be submitted only by FTP or on diskette unless they are encoded properly to submit through e-mail. (DVI files are binary and PostScript files tend to be very large.)

Files sent by electronic mail should be addressed to the Internet address `pub-submit@math.ams.org`. The subject line of the message should include the publication code to identify it as a Memoir. TeX source files, DVI files, and PostScript files can be transferred over the Internet by FTP to the Internet node `e-math.ams.org` (130.44.1.100).

Electronic graphics. Figures may be submitted to the AMS in an electronic format. The AMS recommends that graphics created electronically be saved in Encapsulated PostScript (EPS) format. This includes graphics originated via a graphics application as well as scanned photographs or other computer-generated images.

If the graphics package used does not support EPS output, the graphics file should be saved in one of the standard graphics formats—such as TIFF, PICT, GIF, etc.—rather than in an application-dependent format. Graphics files submitted in an application-dependent format are not likely to be used. No matter what method was used to produce the graphic, it is necessary to provide a paper copy to the AMS.

Authors using graphics packages for the creation of electronic art should also avoid the use of any lines thinner than 0.5 points in width. Many graphics packages allow the user to specify a "hairline" for a very thin line. Hairlines often look acceptable when proofed on a typical laser printer. However, when produced on a high-resolution laser imagesetter, hairlines become nearly invisible and will be lost entirely in the final printing process.

Screens should be set to values between 15% and 85%. Screens which fall outside of this range are too light or too dark to print correctly.

Any inquiries concerning a paper that has been accepted for publication should be sent directly to the Editorial Department, American Mathematical Society, P. O. Box 6248, Providence, RI 02940-6248.

Editors